ELECTROPLATING
with
COBALT

Metals and Alloys Engineering Series

by

Herbert T. Kalmus, Ph.D.

with contributions by

C.H. Harper

and

W.L. Savell

Wexford Press
2008

CONTENTS.

vii

ILLUSTRATIONS.

ELECTRO-PLATING WITH COBALT

INTRODUCTORY.

Investigations of the metal cobalt and its alloys—having as objective the finding of increased commercial uses for them—have been and are being conducted at the School of Mining, Queen's University, Kingston, Ont., under the authority of the Dominion Government, along the following general lines:—

    I. THE PREPARATION OF METALLIC COBALT BY REDUCTION OF THE OXIDE.
    II. A STUDY OF THE PHYSICAL PROPERTIES OF THE METAL COBALT.
    III. ELECTRO-PLATING WITH COBALT AND ITS ALLOYS.
    IV. COBALT ALLOYS OF EXTREME HARDNESS.
    V. COBALT ALLOYS WITH NON-CORROSIVE PROPERTIES.
    VI. THE MAGNETIC PROPERTIES OF COBALT AND OF Fe_2Co.

THE PURPOSE OF THIS INVESTIGATION.

This paper forms part III of the above series, and is a report of researches on electro-plating with cobalt and some of its alloys.

NICKEL PLATING VS. COBALT PLATING.

In 1842, Professor Boetger pointed out that dense and lustrous deposits of nickel could be obtained, which, on account of their resistance to oxidation, great hardness, and elegant appearance, were capable of many applications. The outcome of this has been that during the last decade commercial plating with nickel has developed to be of very great magnitude. On the other hand no plates of cobalt or of its alloys have ever been in extended commercial use. No doubt part of the reason is because of the difficulty of obtaining a supply of cobalt metal at an attractive price. On the other hand, for commercial plating where labour, overhead charges, the preparation of the surface to be plated, the difficulty of maintaining the bath, the amount of buffing, the current efficiency, and particularly the speed with which the work may be run through the baths, are so considerable a fraction of the cost of the finished work—the price of the metal to make up the anodes and the salts is by no means alone the determining factor in the choice of that metal. Moreover, the speed of plating largely determines the hardness and other physical properties of the plate, which in turn determines the weight of metal required for satisfactory commercial tests.

A great many technical points in connexion with the plating of cobalt have not been investigated; the corresponding investigations for nickel have been comparatively thorough. Before platers can adopt cobalt for many purposes, on a considerable scale, a number of questions must be definitely answered by experiments, such as:—

(1) Can cobalt be plated on iron, steel, brass, tin, German silver, lead, etc., to yield as uniform, as adhesive, and as satisfactory a finished surface as nickel?

(2) Is cobalt plate harder than nickel plate?

(3) Is cobalt plate less corrosive than nickel plate under ordinary atmospheric action?

(4) What bath is most suitable for the deposition of cobalt where a heavy protective coating, which may be buffed to a superior finish, is required to be deposited in a minimum of time?

(5) Can a satisfactory cobalt bath be maintained at such an increased concentration as compared with the nickel bath, that plating from it may proceed with greater speed?

(6) Is the cobalt bath more or less troublesome than the nickel bath as regards crystallization, etc?

(7) Should alkali, acid, or neutral baths be used for cobalt plating?

(8) Is the nature of the deposit improved by hardeners such as boric acid, citric acid, magnesium salts, etc?

(9) How does the maximum current density at which cobalt may be deposited commercially, compare with the maximum current densities used in the commercial deposition of nickel?

(10) What electromotive force had best be used for cobalt plating, using the bath found most suitable for a given class of work?

(11) How do cobalt anodes compare with nickel anodes as regards solubility, under the conditions of the plating bath?

(12) What are the relative current efficiencies of cobalt and nickel plating under the best conditions?

(13) How do the electrical conductivities of satisfactory cobalt and nickel plating solutions compare?

(14) Can cobalt be deposited to considerable thicknesses from any solution in accordance with commercial practice?

(15) What is the relative cost of cobalt and nickel plating?

Although numerous statements have been published in the past with regard to cobalt plating, the conclusions to be drawn from a survey of the existing literature and patents would lead one to be very sceptical as to the advantages of cobalt plating over nickel plating. It is noticeable, however, that the conditions for the production of good cobalt plates, as given by different authors, vary very greatly among themselves. Not only are the conclusions often diametrically opposite, but likewise the data from which these conclusions are drawn.

Consider alone the question of the relative maximum current densities with which cobalt and nickel may be successfully plated. There is little or nothing in the literature relating to the solutions of cobalt which we have found in this laboratory to be most suitable for fast plating. If it can be shown to be practically feasible to plate cobalt from a bath at several times the speed that this is possible for nickel, other things being equal or in favour of the cobalt, this greater speed of turning out the work, with attendant economies, might offset a very considerable difference in the initial cost of the anodes and salts of the two metals. It must certainly appeal to anyone that if cobalt-ammonium sulphate, because of its very much higher solubility than nickel-ammonium sulphate, or for other reasons, may be used as a plating bath at very much higher current densities, that with such a bath the plater may turn out work at correspondingly greater speed. Moreover, plating at greater speed will probably mean a harder and denser plate with consequent less weight of metal required.

A large number of plating experiments were undertaken at this laboratory with a view to studying the questions outlined above. Numerous types of cobalt baths were used and experiments performed with various concentrations of each solution. The object of this paper is to set forth the data of these experiments and the conclusions to be drawn therefrom.

The following paragraph shows briefly the wide variance in opinion to be found in reviewing the most important contributions to this subject in the literature.

LITERATURE ON ELECTRO-DEPOSITION OF COBALT.

Langbein, in his well-known work, "Electro-deposition of Metals,"[1] devotes nearly one hundred and seventy-five pages to the deposition of nickel, and a scant two pages to the deposition of cobalt. With regard to cobalt, Langbein remarks as follows[1]:—

For plating with cobalt, the baths given under "Nickelling" may be used by substituting for the nickel salt a corresponding quantity of cobalt salt. By observing the rules given for nickelling, the operation proceeds with ease. Anodes of metallic cobalt are to be used in place of nickel anodes.

Nickel being cheaper and its colour somewhat whiter, electroplating with cobalt is but little practised. On account of the greater solubility of cobalt in dilute sulphuric acid, it is, however, under all circumstances, to be preferred for facing valuable copper plates for printing.

According to the more or less careful adjustment of such plates in the press, the facing in some places is more or less attacked, and it may be desired to remove the coating and make a fresh deposit. For this purpose Gaiffe has proposed the use of cobalt in place of nickel, because the former dissolves slowly but completely in dilute sulphuric acid. He recommends a solution of one part of chloride of cobalt in ten of water. The solution is to be neutralized with aqua ammonia, and the plates are to be electroplated with the use of a moderate current.

From our experiments, we find in general that the substitution of cobalt for nickel in Langbein's baths, as recommended by him, does not do justice to the metal cobalt. These baths have been primarily worked out for nickel, whereas cobalt, being in many instances of far greater solubility, required a much more concentrated bath than that recommended, for the fastest and best plating. Some of our best cobalt solutions are not included at all in Langbein's list.

MacMillan,[2] Watt,[3] and others, report that cobalt deposits are harder than the corresponding nickel deposits. Brochet,[4] says:—

Cobalting has been proposed in place of nickelling when a deposit of greatest hardness is desired.

On the other hand, Langbein,[5] Wahl,[6] and others, say that cobalt plates are softer than the corresponding nickel ones.

S. P. Thompson[7] finds that articles plated with cobalt are less corroded in the atmosphere of London than either silver or nickel plate, while Stolba[8] reports that cobalt salts treated like nickel salts yield metallic deposits of a steel grey colour, less lustrous than nickel and more liable to tarnish.

[1] "Electro-Deposition of Metals," Langbein, 6th Edition Revised, Henry Carey Baird and Co., Philadelphia, p. 318.
[2] Electro-Metallurgy, MacMillan, 1901, p. 227.
[3] Electro-plating and Electro-refining of Metals. Watt-Phillips, 1902, p. 360.
[4] Manuel Pratique de Galvanoplastie, 1908, p. 313.
[5] Electro-plating of Metals, 6th Edition, p. 319.
[6] Electro-plating and Electro-refining of Metals, Watt-Phillips, 1902, p. 360.
[7] Journal of Institute of Electrical Engineers, 1892, p. 561.
[8] Electrical Review, Nov. 18th, 1887, p. 503.

While some authorities report that it is practically impossible to obtain a deposit of cobalt more than a few hundredths of a millimetre in thickness, Bouant[1] says:—

Electrolytic deposits of cobalt are easily obtained even of a very great thickness, so that electro-deposition of cobalt is as easy as that of copper.

A review of much of the literature on the electro-deposition of cobalt was presented at the meeting of the American Electro-chemical Society, at Atlantic City in April, 1913, by O. P. Watts. A consideration of Watts' paper emphasizes the diversity of opinion with regard to the cobalt plating mentioned above.

Watts mentions about fifty solutions of cobalt which have been used with greater or lesser success. Some of these are similar to those used by us in our experiments, but several of our best solutions are not included in the list.

Many of these baths are recommended only for very weak currents; others are more or less indefinite, and there is much contradiction among them. For example, the first solution in Watts' list is mentioned as follows:—

Becquerel's solution, 37·5 grams of cobalt chloride neutralized by ammonia or potassium hydroxide, gives a brilliant, white, hard and brittle deposit, a very weak current must be used.

Later, quoting from Watts, Isaac Adams says, in a patent application:—

I have found that a solution made and used in the manner described in the books will not produce such a continuous and uniform deposit of cobalt as is necessary for the successful and practical electroplating of metals with cobalt... I have found further that the simple salts of cobalt, such as are recommended by Becquerel and others, are not such salts as could be used in practical electroplating with cobalt . . . . . .

Even in a matter so immediate and important to the practical plater as the acidity or basicity of his plating bath, there is no general agreement among the authorities even for nickel solutions, much less for cobalt. Dr. Langbein[2] writes:—

All nickel baths work best when they possess a neutral or slightly acid reaction . . . an alkaline reaction of nickel baths is absolutely detrimental, such baths deposit a metal dull and with yellowish colour, and do not yield thick deposits. Bennett, Kenny and Dugliss, in a paper read before the American Electro-Chemical Society at New York City, April, 1914, state as two of their seven general conclusions the following,—"A good deposit of nickel may be obtained from the double sulphate if the solution at the surface of the cathode is kept alkaline . . . . since alkalinity is necessary for good efficiency, it is very probable that in acid solutions nickel is deposited only when impoverishment of the hydrogen ions has caused the solution to become alkaline and given the conditions under which nickel may deposit.

The foregoing are but a few of many contradictory statements to be found in the literature on electro-deposition of cobalt.

SUBDIVISION AND ARRANGEMENT OF EXPERIMENTS ON ELECTRO-DEPOSITION OF COBALT.

A very large number of plating experiments were conducted by us at this laboratory for the purposes already indicated; in connexion with

---

[1] La Galvanoplastie, Bouant, 1894, p. 294.
[2] Electro-Deposition of Metals, Dr. Geo. Langbein, 6th Edition Revised .p. 251.

which, some sixteen different types of solution or baths were employed and studied.   The following is a list of the series:—

Series   (1)   Simple cobalt-ammonium sulphate.
"   (2)   Cobalt-ammonium sulphate, with an excess of ammonium sulphate.
"   (3)   Cobalt-ammonium sulphate, with an excess of ammonium sulphate to which is added citric acid.
"   (4)   Cobalt-ammonium sulphate, with ammonium chloride.
"   (5)   Cobalt chloride with ammonium chloride.
"   (6)   Cobalt-ammonium sulphate with boric acid.
"   (7)   Cobalt-ammonium sulphate, cobalt carbonate and boric acid.
"   (8)   Cobalt sulphate, potassium citrate and ammonium chloride.
"   (9)   Cobalt phosphate with sodium-pyrophosphate.
"   (10) Cobalt-ammonium sulphate with magnesium sulphate.
"   (11) Cobalt sulphate, neutral ammonium tartrate, with the addition of tannic acid.
"   (12) Cobalt sulphate, potassium tartrate and tartaric acid.
"   (13) Cobalt sulphate, sodium chloride and boric acid.
"   (14) Cobalt sulphate, ammonium sulphate, magnesium sulphate with boric acid.
"   (15) Cobalt-ethyl sulphate, sodium sulphate and ammonium chloride.
"   (16) Cobalt sulphate, ammonium sulphate, ammonium chloride and boric acid.

Each set of plating experiments with a definite group of principal components in the bath, has been given a series number. Variations in concentration of this bath, or changes in the relative proportions of the components in the bath, have been designated by the use of subscript letters, A, B, and C, etc. This subscript notation has also been used to include minor changes and additions to a bath. Also, where a nickel bath has been used, analogous to a given cobalt bath, the data for the experiments on it are noted under the corresponding cobalt series. For example, Series 3 refers to baths of cobalt sulphate, ammonium sulphate, and citric acid. A combination of these was called III A; whereas the same bath, with a small amount of sodium sulphite, to note its effect, was called III B, and the same bath, substituting 34·5 grams of nickel sulphate for the 34·5 grams of cobalt sulphate in solution III A, was called solution III C. In this way all the modifications of a bath are grouped under one series.

METHOD AND ARRANGEMENTS FOR PLATING EXPERIMENTS.

ARRANGEMENT OF ELECTRICAL CIRCUIT.

Power was taken from three storage batteries connected in series, yielding, approximately, 6 volts. The various baths were connected independently across the 6 volt terminals of this storage battery set, with appropriate resistance in series with each. In this way the potential across the electrodes of each was cut down to the desired value. A five volt range Weston voltmeter, reading to 0·01 volts, was connected directly across the electrodes of each bath. A Weston milli-ammeter, 600 milli-ampere scale, was connected in series with each bath.

2

PLATING TANK, AND COMPUTATION OF EFFECTIVE ELECTRODE SURFACE

The plating tanks used for these experiments were of glass, rectangular in shape, and of approximately the following internal dimensions, 12″ × 6″ × 6″: some of them were deeper than this.

In general, if the bath is not large in cross-section in comparison with the area of the electrodes, the results obtained will not be reproducible. We tried experiments using a tank not much larger in cross-section than the electrodes. In that case the voltage required for a given current density, with electrodes 10 cms. apart, is very much greater than with a tank of considerably larger cross-section, other conditions being the same. This is largely due to the fact that, with the smaller cross-section of tank the available solution between the electrodes is diminished.

For the most part, our electrodes were approximately 3″ × 2″ in surface, placed from 10 to 20 cms. apart, in the centre of the rectangular plating tank above mentioned. If the electrodes be kept in place, the tank lowered away from them, and substituted by a very large tank of the same solution, we found that the electromotive force across the electrodes for a given current density diminished by nearly 20 per cent. The leads of the electrodes were coated with a layer of insulating asphaltum, as described in the paragraph on electrodes. In the case of these experiments, therefore, the effective electrode area is something like 20 per cent less than the exposed surface.

Again, instead of submerging electrodes of the above size in the middle of the tank, in some of our experiments the electrodes were lowered into a tank to such a depth as to be just covered by the solution. It is obvious that in this case, as we found by test, the electromotive force across the electrodes would diminish if the height of the bath around them be raised. This is true even though no greater area of electrode is thereby covered with solution, and although the leads are insulated with asphaltum. The reason is, that with the level of solution just covering the top of the electrodes, the electrical field between the upper portion of the electrodes was limited by the surface of electrolyte, as compared with what it would be with the electrodes immersed in the centre of a large body of electrolyte.

Consequently, had our experiments been performed in an ideally large tank, the electromotive forces would have been smaller for the same current densities. We preferred to use a bath approximating to the conditions of plating practice, and have in all cases noted in the tables to follow, the values of current density, electromotive force, etc., just as observed without correction.

Good practice for commercial nickel plating with many solutions is, to have about 8 gallons of solution for each square foot of anode surface. In our experiments we have not departed widely from this value, although we find different solutions require a somewhat different magnitude of this ratio for the best results.

THE PLATING EXPERIMENTS.

The plating experiment itself consisted merely in noting changes in appearance of the solution and electrodes as the deposition progressed, as well as making a record of millimetre, voltmeter, and time readings at frequent intervals. The concentration of the bath was measured from time to time, and the physical properties of the resulting plate were studied.

Wherever current efficiencies were desired, millimetre readings were made at intervals of not more than two minutes. These readings were very constant in all cases, so that more frequent readings were not necessary.

Also the cathode was carefully dried and weighed before and after the deposition. Precautions were taken so that conditions of weighing the electrode before plating and after plating were identical.

Advantage was taken of the well-known fact for the deposition of nickel and other metals, that the deposit is more adhesive when struck on initially at a higher potential than is to be used throughout the greater part of the plating run. Our practice was to diminish the resistance in series with the bath, so that an initial electromotive force of about six volts was used, sending very considerable currents through the bath during approximately the first thirty seconds of the run. The effect of this was uniformly satisfactory in causing the plates to adhere firmly.

After a sufficient plate had been deposited on the cathode, it was removed from the bath, immediately washed in cold water, and then rinsed in boiling water until it came to a uniform temperature, after which it was dried in hot sawdust. This procedure was practised throughout the plating experiments.

BUFFING AND FINISHING COBALT PLATES.

The cathode after being removed from the plating bath and dried, is in general, white and metallic, and requires but little buffing to produce a satisfactory mirror surface. Our practice has been to cut the plate slightly with tripoli and rag wheels, and then to "colour" the surface with soft buffs, as Vienna lime, to a high finish.

PREPARATION AND USE OF ELECTRODES.

ANODES.

Both cast and rolled anodes were used for these experiments, the cobalt employed analyzing as follows:—

		%
H 218	Co	95·70
	Ni	0·80
2 Anodes	Fe	2·10
	S	0·043
$3\frac{1}{2}'' \times 2'' \times \frac{1}{4}''$	C	0·040
	P	0·004
Cast	Si	0·050

		%
H 219	Co	98·00
	Ni	0·75
4 Anodes	Fe	1·35
	S	0·042
$3\frac{1}{2}'' \times 2'' \times \frac{1}{4}''$	C	0·060
	P	0·007
Cast	Si	0·067

		%
H 209	Co	98·75
	Ni	none
6 Anodes	Fe	1·35
	As	none

		%
$7'' \times 2\frac{1}{2}''$ sent to Russel Motor Car Co. for	P	0·0067
experiments.........................S	S	0·052
	C	0·061

		%
H 221..............................Co	Co	95·63
	Ni	0·75
4 Anodes.............................Fe	Fe	3·72
	P	0·005
$3\frac{1}{2}'' \times 2'' \times \frac{1}{4}''$.............................S	S	0·029
	C	0·052
Rolled...............................As	As	0·048

The cast anodes were poured in sand and smoothed down with an abrasive wheel to a finished surface of approximately $3\frac{1}{2}'' \times 2'' \times \frac{1}{4}''$.

The rolled anodes were made from ingots about $8''$ in length, and 1 sq. inch in cross-section, which were rolled down to plates $\frac{1}{4}''$ thick, from which anodes $3\frac{1}{2}'' \times 2'' \times \frac{1}{4}''$ were finished.

Impurities in Anodes.

Nickel anodes, as sold in the market, frequently run in the neighbourhood of 92 per cent nickel and $7\frac{1}{2}$ per cent to 8 per cent of iron. The iron is cast in the anode because pure nickel is not corroded rapidly enough under most conditions to furnish the necessary metal to the bath. Iron is a cheap material, and has a solution tension enough greater than nickel to make it effective for the end in view. The greater solution tension of cobalt in the plating baths, as compared with nickel, renders practicable the use of a lesser amount of iron, or of no iron at all. Very pure cobalt anodes were readily dissolved in the solution under the conditions of many of our best plating tests.

The freedom of the cobalt anode from iron no doubt has much to do with the diminished corrosion of the cobalt plate as compared with the nickel plates. This will be discussed subsequently in the text.

Rolled versus Cast Anodes.

Cast anodes of both nickel and cobalt go into solution much more readily than the rolled anodes. With a number of each in a nickel or cobalt bath, the proportion between the two may be so chosen that the composition of the electrolyte remains constant as the anode goes into solution. If too large a proportion of cast anodes is used, the anode dissolves with too great readiness, and the bath may become alkaline. Conversely, if too large a proportion of rolled anodes is used, the solution of the anode may not take place with sufficient readiness, and the bath may become acid, and depleted in metal. The greater solubility of cobalt anodes in a number of the plating baths described in this paper, as compared with nickel anodes in the corresponding nickel bath, renders the use of a larger proportion of rolled anodes possible in the case of cobalt than is customary at present with nickel in the nickel plating trade. The greater solubility of the cobalt anode is distinctly in its favour for practical plating purposes.

CATHODES.

The cathodes for these experiments were of brass, iron, and steel, and were usually $3\frac{1}{2}'' \times 2'' \times \frac{1}{4}''$ in dimension. One side of the cathode was given a high polish with emery and buffing wheels, and the other was covered with a thin coat of asphaltum varnish.

It is of the utmost importance in all plating work that the cathode be absolutely smooth, and thoroughly cleansed from particles of dust, grease, etc. This was accomplished by the following procedure. Brass, steel, or other stock, as required, was first machined down to the required size, after which they were smoothed down with an emery wheel, and finished with an appropriate buffing wheel to a mirror surface. This left the surface with a certain amount of grease, and adhering buffing material, which was removed by scrubbing with Kalye caustic solution. After thorough rinsing with water, the electrode was immersed in dilute hydrochloric acid, or in dilute potassium cyanide solution and again thoroughly rinsed with water. It was then ready for use in the bath.

In addition to smooth plane cathodes, a number of them were prepared by cutting patterns, and depressions on one side. These were used in the same manner as the smooth ones after cleansing and buffing as described.

PREPARATION OF SALTS.

The ammonium sulphate, ammonium chloride, boric acid, citric acid, potassium citrate, tannic acid, sodium sulphite, sodium phosphate, ammonium tartrate, and magnesium sulphate, used for these experiments, were Merck's chemically pure, purchased from a supply house; as were the reagents used in the preparation of the cobalt salts.

All cobalt compounds used for these experiments were prepared at Queens laboratory, and the method of preparation of each is given as follows:—

COBALT SULPHATE.

The cobalt sulphate used for these experiments was prepared by dissolving Co_3O_4 in HCl, and adding an equivalent of H_2SO_4. The solution was then evaporated to drive off HCl and crystallize $CoSO_4 \cdot 7H_2O$. The crystals were collected in the usual manner and heated in the presence of H_2SO_4 until SO_3 fumes disappeared and the salt became dehydrated. Or the collected crystals were re-crystallized in the usual manner where dehydrated salt was not desired.

COBALT AMMONIUM SULPHATE.

This double salt was prepared by mixing hot solutions of cobalt sulphate and ammonium sulphate containing molecular proportions of each of the salts. As the cobalt ammonium sulphate is less soluble than either of the single salts, most of the crystallization was effected in the hot mixed solution. The crystals were collected and washed with alcohol.

COBALT CHLORIDE.

The cobalt chloride was prepared by dissolving Co_3O_4 in HCl, and after filtering off the excess Co_3O_4 crystallizing the salt and collecting and washing the crystals in the usual manner.

COBALT ETHYL SULPHATE.

Ethyl alcohol and concentrated sulphuric acid were heated together to form ethyl sulphuric acid. This was then neutralized with calcium carbonate, and the calcium ethyl sulphate crystallized by evaporation of the solution after filtering off precipitated calcium sulphate. The crystals were re-dissolved and the calcium precipitated as calcium sulphate by adding sulphuric acid in molecular proportion, reforming ethyl sulphuric acid. Freshly precipitated cobalt hydroxide was added to this acid until it failed to dissolve any more. The cobalt ethyl sulphate was crystallized by evaporation. The crystals were collected and washed in the usual manner.

THE SOLUBILITY OF COBALT AND NICKEL SALTS.

The relative solubilities of cobalt sulphate and nickel sulphate, and of cobalt-ammonium-sulphate and nickel-ammonium-sulphate are important in considering the greater conductivities of the cobalt solutions as compared with the corresponding solutions of nickel. These solubilities are considered later when discussing the greater speed of cobalt plating, and as well in considering the saturation concentration of some of the baths employed, so that a brief table of solubilities at room temperature is given herewith.

Solubility Table.

Grams of salt dissolved by one litre of water.

			Comey[1]	Kraut[2]	This Laboratory
$CoSO_4$................anhydrous	23°C	362	380	362·2
$Ni\ SO_4$................	"	"	405	379	363
$Co\ (NH_4)_2\ (SO_4)_2$......	"	"	171	171	164
$Ni\ (NH_4)_2\ (SO_4)_2$........	"	"	66	66	72·8

The salts actually used in making up the solution were $Co\ SO_4 \cdot 7H_2O$, $NiSO_4 \cdot 7H_2O$, $CoSO_4 \cdot (NH_4)_2\ SO_4 \cdot 6H_2O$ and $Ni\ SO_4 \cdot (NH_4)_2\ SO_4 \cdot 6H_2O$.

From the above table it is apparent that the solubilities of cobalt and nickel sulphate are not very different. These, however, do not lend themselves to satisfactory plating solutions without adding various other compounds to them. On the other hand, cobalt ammonium sulphate is approximately 2·5 times as soluble as nickel ammonium sulphate. As will be shown in the sequel, from the experiments performed, the saturated solution of this cobalt ammonium sulphate offers an extremely rapid and satisfactory plating bath.

CURRENT DENSITIES

Unless specifically stated to the contrary, the current densities given in the tables to follow and throughout the text are cathode current densities. That is, they were computed by dividing the total current by the cathode area.

An approximate notion of the anode current density may be obtained from the following statement, although it is impossible to compute it with any degree of accuracy. At the beginning of each experiment, the anode was smooth and of approximately the same area as the cathode, see p. 8, but as the experiment proceeded, the anode gradually increased in effective area, due to unequal corrosion of its surface. At the end of some of our experiments the anode area may have been more than twice that of the cathode.

---

[1] A dictionary of Chemical Solubilities, A. M. Comey—MacMillan and Co., 1896.
[2] Gmelin—Kraut's Handbuch der anorganischen Chemie, 1909, Vol. V, Sect. 1.

In our conclusions and through the text of this paper we speak of allowable current densities with the various solutions, and we recommend certain ranges of current density which are permissible for the best plates. In every case these are cathode current densities, but are, of course, only valid with a sufficient anode area. We have conducted our experiments under conditions which might be called good standard practice for plating shops, which means, so far as anode area is concerned, that the anode area was in every case greater than, and in many cases, twice that of the cathode area.

AGEING TESTS

With all the plating baths which were found to be satisfactory or promising, both ageing and efficiency tests were run. The ageing tests were for the purpose of ascertaining the constancy of the bath during the continued use for plating over a considerably longer period than was necessary for our experiments. The solution was analyzed for cobalt content, and tested for acidity and alkalinity at intervals of 15 hours during plating runs of one hundred or more hours' duration. This was in addition to the series of runs already made with the given bath, operating with it to produce the plates studied. The results of these ageing tests are given with each set of experiments.

EFFICIENCY TESTS

Efficiency tests were made in the usual manner by carefully weighing the quantity of metal deposited in a measured time and comparing it with the theoretical amount of metal which should have been deposited in accordance with Faraday's laws. The data for these tests are given for each important solution in its appropriate place after the plating and ageing tests.

EXPERIMENTS: ELECTRO-DEPOSITION OF COBALT.

Following, are described the various solutions which were used for the cobalt plating experiments, together with the data of the plating experiments themselves. After each series are recorded the conclusions which were drawn therefrom.

SERIES 1: COBALT AMMONIUM SULPHATE SOLUTIONS.

SOLUTION I A.

This solution was analogous to one recommended by Dr. Langbein for nickel plating, and consisted of:—

Cobalt sulphate	$CoSO_4$	30·9 grams.
Ammonium "	$(NH_4)_2SO_4$	26·3 "
Water		1000 c.c.

The total amount of the bath prepared was 3 litres.

This bath contained cobalt sulphate and ammonium sulphate in molecular proportions, and by reference to page 10, it is obvious that it is only somewhat more than one-third saturated.

Comparison of Nickel and Cobalt Solutions of the Type of Series 1.

Langbein recommends for the nickel bath of the type I A, with an electrode distance of 10 cms, that an electromotive force of 3 volts be employed, and he states this will give a current density of 0·3 amperes per square decimetre.[1]

We wished to compare the cobalt solution I A with the corresponding nickel solution as regards their relative specific electrical conductivities or with regard to the electromotive force necessary to give a specific current density through each, other conditions being the same. An experiment was tried using solution I A, with which was compared the following nickel-ammonium sulphate solution:—

Nickel sulphate $NiSO$................ 30·9 grams.
Ammonium sulphate $(NH_4)_2SO_4$........... 26·3   "
Water................................. 1000 c.c.

Using the proper current to obtain a current density of 0·3 amperes per square decimetre, with the cobalt solution I A, it was found that the potential differenec between the electrodes was 0·88 volts. With the identical electrodes and with the same distance between them, namely 10 cms. using the nickel solution, the potential difference between the electrodes was 2·1 volts. Thus we have the very great difference between 0·88 volts and 2·1 volts required to give the same current density with these two solutions, other conditions being equal, which means that the cobalt solution is of very much lower specific electrical resistance than the nickel solution.

Thus, in the first series of experiments with cobalt-ammonium sulphate and nickel-ammonium sulphate, we found a result which ran uniformly throughout our plating experiments, namely, that the conductivity of the cobalt bath was considerably higher than that of the corresponding nickel bath.

---

[1] Dr. Geo. Langbein, "Electro-Deposition of Metals," 6th Edition, Revised, P. 252.

Data of Plating Experiments, Solution, I A.

Date of experiment.	Current density in amperes per sq. decimetre.	Electromotive force in volts.	Electrode distance in cms.	Cathode area in sq. cms.	Duration of plating.		Character of plate and remarks.

Cast Anodes.

Date of experiment.	Current density	Electromotive force	Electrode distance	Cathode area	hr.	min.	Character of plate and remarks.
1914 June 8	0·30	2·45	46	Brass 27·3	2	30	Deposit fairly bright and hard but pitted at bottom.
June 8	0·30	2·75	45	27·3	1	0	Deposit light grey in colour and somewhat pitted at bottom.
June 22	0·30	0·75	10	27·3	2	0	Plate bright and smooth, but not lustrous.
June 25	0·45	0·83	10	27·3	4	0	Heavy metallic lustrous plate slightly spotted.
June 27	0·50	0·80	10	27·3	3	0	Plate metallic and lustrous but streaked and spotted.
June 24	0·60	1·06	10	27·3	2	0	A metallic uniform though not lustrous deposit.
June 25	0·80	1·27	10	27·3	3	0	Plate uniform, metallic and lustrous.
June 27	0·90	2·03	20	27·3	1	30	Plate bright and smooth except at bottom which was rough but not burned.
June 26	1·0	1·52	10	27·3	1	30	A uniform plate except at lower corners which were burned. Dull in colour.
June 30	1·0	2·25	20	27·3	1	30	Plate uniform and extremely bright, giving a thoroughly satisfactory surface when polished.
July 3	1·0	2·45	20	27·3	2	30	A uniform bright plate except at edges which appeared slightly burned. Solution boiled just prior to this run.
July 23	1·25	2·00	10	27·3	1	15	Deposit dark and badly burned at edges of cathode, bright in centre.
July 23	1·50	2·10	10	27·3	1	15	Same as preceding, but burning more marked.

Rolled Anodes.

Date	Current density	Electromotive force	Electrode distance	Cathode area	hr.	min.	Character of plate and remarks.
July 17	0·50	0·83	10	Zinc 30·1	1	0	Plate showed very rough surface, unlike anything obtained on brass or steel. Under microscope metal appeared to be deposited in small irregular masses and not to be polished.
July 16	0·80	2·0	10	Brass 37·0	1	0	Exceptionally good white, uniform plate.
July 17	1·0	2·0	17	Brass 37·0	1	0	Plate slightly burnt at upper edges.
July 17	1·0	2·75	17	Brass 37·1	1	0	A clean, smooth, bright plate as removed from the bath, which peeled, however, upon heating in boiling water for a few minutes.

Ageing Test, Solution I A.

	Grams cobalt in 100 c.c. solution	Acidity of solution
Solution at end of above series of experiments, Aug. 26, 1914, after 31 hours plating as shown	0·94	Neutral
Same after 46 hours plating	0·94	"
" " 61 " "	0·95	"
After 76 hours plating solution was diluted slightly to original volume to make up for evaporation	0·94	"
After 87 hours plating	0·96	"
" 102 " "	0·97	"
" 117 " "	0·99	"

The last 90 hours plating were with a current density of 0·60 amperes per square decimetre.

It is, therefore, clear that this solution is not changing rapidly, and that the anode is dissolving satisfactorily. The slight increase in concentration is accounted for by evaporation.

Very similar results were obtained with iron and steel cathodes in place of brass.

Solution I A was not thought sufficiently important to warrant a similar series of experiments being made with the corresponding nickel bath for comparison.

SOLUTION I B.

This solution is a nearly saturated solution of cobalt-ammonium sulphate and was made by dissolving 200 grams $CoSO_4 (NH_4)_2 SO_4 \cdot 6H_2O$ to the litre of water, which is equivalent to 145 grams $CoSO_4 \cdot (NH_4)_2 SO_4$ to the litre of water.

No reagent was added to this solution after it was made up to maintain its neutral reaction or for other purposes.

Specific Gravity = 1·053 at 15°C.

Data of Plating Experiments, Solution I B.

				Cast Anodes.		
Date of experiment.	Current density in amperes per sq. decimetre.	Electromotive force in volts.	Electrode distance in cms.	Cathode area in sq. cms.	Duration of plating.	Character of plate and remarks.
1914 July 17	0·75	1·0	17	Brass 37·1	hr. 1 min. 30	Bright, uniform glossy plate, which required but little buffing to give a satisfactory surface.
July 17	1·0	1·5	19	Brass 37·1	1 30	Same
July 21	1·0	1·5	24	Brass with grooved channels to afford high and low spots, 23·7	1 0	A good smooth bright plate obtained over entire surface. When cobalt plate was dissolved off in nitric acid, grooves lost their plate first showing thinner deposit there than on higher places.
Aug. 3	1·0	1·70	19	Brass 23·7	2 0	Very satisfactory plate, hard, bright and easily buffed.
July 18	1·2	1·95	19	Brass 37·1	1 0	Bright uniform glossy plate which required but little buffing to give a satisfactory surface.
July 22	1·2	1·7	24	Brass with grooved channels to afford high and low spots. 23·7	1 0	A beautiful smooth plate obtained over entire surface. High and low spots evenly coated.
July 22	1·5	2·05	24	Same	2 0	This plate was very smooth and bright on removal from bath, and after slight buffing, assumed a beautiful finish. Uniform in thickness on high and low spots. Thickness of plate 0·062 mms.
July 30	1·5	1·20	10	Polished steel knife	1 30	A bright, smooth hard even satisfactory deposit.
Aug. 5	1·5	2·1	16	37·1	1 30	Beautiful bright smooth plate easily buffed to mirror surface.
July 22	1·7	2·4	24	Brass with grooved channels to afford high and low spots 23·7	1 15	This plate was very smooth and bright on removal from bath and after slight buffing assumed a beautiful finish. Uniform in thickness on high and low spots.

Cast Anodes— Continued

Date of experiment.	Current density in amperes per sq. decimetre.	Electro-motive force in volts.	Electrode distance in cms.	Cathode area in sq. cms.	Duration of plating.		Character of plate and remarks.
July 23	1·8	2·55	24	Same	2	45	Same
July 24	2·0	2·85	24	Same Area 28·0	2	15	A beautiful white, hard lustrous plate obtained with no sign of burning or scaling.
July 28	2·2	3·35	24	Same Area 27·3	1	30	Same
July 28	2·5	3·75	24	25·0	2	45	A very smooth even hard plate, white in colour which, after slightly buffing, assumed beautiful satisfactory finish.
July 28	2·7	2·25	23	22·2	1	15	Same. No signs of burning.
July 31	3·5	3·0	22	Brass with grooved channels to afford high and low spots 20·0	1	30	Very smooth even hard plate white in colour, which after very slightly buffing, assumed beautiful satisfactory finish. No signs of burning.
July 31	4·0	2·9	20	Same 18	1	0	Same
Aug. 3	4·5	4·9	20	Brass 27·5	1	30	Plate fairly uniform and bright, but darkened at edges. Began to peel at one corner.
Aug. 3	4·5	3·0	10	Brass 27·5	1	30	Plate similar to last experiment with burning slightly marked, but no peeling.
Aug. 3	4·5	2·95	10	Brass 27·5	1	0	Plate similar to previous ones, but burning not quite so marked. Buffed satisfactorily, but showed slightly pitted due to gas bubbles.

Rolled Anodes.

Date of experiment.	Current density in amperes per sq. decimetre.	Electro-motive force in volts.	Electrode distance in cms.	Cathode area in sq. cms.	Duration of plating.		Character of plate and remarks.
1914 Sept. 1	0·39	0·44	10	Brass 26·2	2	0	Smooth uniform white plate, readily buffed to mirror surface.
Aug. 31	0·61	0·55	10	Brass 26·7	2	0	Same
Aug. 5	1·0	1·6	16	Brass 37·0	24	0	On removal from solution plate was very bright and metallic, but had split at one place and could be easily removed from the brass cathode. Thickness of plate 0·34 mm. at edge, 0·24 mms. at centre.
Sept. 2	1·0	0·77	10	Brass 27·0	1	0	Smooth uniform white plate, readily buffed to mirror surface.
Aug. 4	1·5	2·2	16	Brass 37	16	0	Beautiful bright smooth plate, which began to separate from cathode at one corner.
Sept. 14	2·0	1·30	10	Brass 26·5	0	45	Very even smooth deposit, having a good lustre when polished.
Sept. 15	3·0	1·32	10	Brass 20·0	0	20	White plate, good lustre when polished, somewhat rough on bottom, showing heavier deposit there, also small furrows from gas streaks.
Sept. 14	4·0	1·35	10	Brass 11·7	0	15	Splendid white deposit, but cracked a short distance in one place.
Sept. 15	4·0	1·35	10	Brass 11·7	0	25	Good white plate, but rough on bottom, showing heavier deposit there, also very small furrows from gas streaks.
Aug. 4	4·25	4·25	20	Brass 27	1	15	Burned but not so badly as 4·5 ampere current density.
Aug. 4	4·5	4·60	18	27	1	0	Plates burned along edges.

A number of the above plates were given a very severe bending test to study their adhesive qualities. In every case the plates stuck to the cathode after being bent backwards and forwards at an angle of nearly 180 degrees in a manner equal to, if not better than that of the best nickel plates with which we are familiar, subjected to a similar test. Even after the surface of the metal base had started to break, the cobalt plate still continued to cover the furrows and ridges formed.

Further Data of Plating Experiments with Cobalt Solution I B.

Experiments at Russell Motor Car Company Plating Plant.

Solution: 5 lbs. salts, 6 gal. water, Sp. Gr. 1,050—acidity, neutral.

Date	Current density in amps. per sq. dec.	Electromotive force in volts.	Electrode distance in inches.	Cathode area in sq. inches.	Duration of plating.	Anode area in sq. inches.	Remarks.
					hr. min		
Sept. 10	8·6	3·75	5	Steel 15	0 ½	28	Blackened in 30 seconds.
Sept. 10	3·4	3·25	5	Steel 45	0 40	56	Poor plate, buffed easily, but blisters appeared, long and narrow in form. small pinholes on some portions, Slightly burned on lower ends.
Sept. 11	2·3	1·25	5	Brass Hub cap 15	0 5	28	Deposit slow, spotted, streaked, very dark.
Sept. 11	4·3	2·0	5	Same 15	0 5	90	Deposition rapid. Plate spotted, not uniform. Cathode not covered at end of run.
Sept. 12	3·9	2·5	5	Same 15	0 5	90	Same
Sept. 12	3·3	3	5	Steel 10	0 30	28	Good, firm adherent plate. Very hard and smooth. Colour similar to nickel before buffing. Coloured easily and to good finish of slightly bluish tone.
Sept. 12	3·5	2·75	5	Steel 36	0 15	90	Did not cover, streaked, very hard brittle plate.
Sept. 12	3·3	3	5	Brass 15	0 50	28	Good colour. Porous spots in casting refused to cover at edges. Indented portions of cathode not coated. Spot beneath slinging wire bare.
Sept. 14	3·0	2·5	5	Brass 15	1 0	28	Struck the piece with nickel for 30 seconds. Cobalt deposit began readily and uniformly. Colour darkened quickly, imperfect in spots, rough and streaked.
Sept. 14	3·0	2·5	5	Steel 12	0 15	28	Deeper portions of piece were not covered.
Sept. 14	2·3	2	5	Steel 4 (Solution rendered slightly alkaline)	0 5	14	Plate scaled, chipped. Deeper portions not covered.
Sept. 14	3·0	2·5	5	Steel 12 (Solution rendered slightly acid)	0 15	28	Plate peeled badly. Several tests were made with similar results.
Sept. 15	1·6	1·5	5	Steel 40	0 15	42	Colour very dark, otherwise a good plate, buffed easily did not cut through.
Sept. 15	1·0	1·5	5	Steel 40	0 5	42	Colour very dark, buffed to good finish.
Sept. 15	2·3	1·5	5	Steel 36	1 0	42	Surface covered instantly and in good condition. Gas liberated more freely than heretofore, but no indication of pinholes or streaks. Colour very dark, resembling burnt nickel deposit. When dried it had a velvety appearance. Soft to the touch, buffed easily on soft wheel.

Continued

Date	Current density in amps. per sq. dec.	Electro-motive force in volts.	Elect-rode distance in inches.	Cathode area in sq. inches.	Dur-ation of plating.	Anode area in sq. inches.	Remarks.
					hr. min.		
Sept. 15	3·0	2	5	Brass 45	0 30	56	Very dark. Surface covered with minute blisters or pits, which disappeared when buffed. Surface covered quickly and completely. Colour satisfactory after buffing.
Sept. 15	3·0	2	5	Steel 40	0 30	56	Very dark, buffed easily and to good finish. Very hard plate, superior to 1 hour nickel from Prometheus.
Sept. 16	...	....	..	..	..	..	Tests made during the following ten days resulted in similar plates, 6 lbs. cobalt salts were then added to the bath. Sp. Gr. 12° Be. No further additions were made—acidity, very slightly acid. After ageing treatment the solution was practically neutral.
Sept. 28	2·1	1·75	5	Steel 36	0 25	42	Surface covered well, gas not as free as formerly, deposit marred by circular mark entire length and breadth of cathode. Colour darker than nickel, hard and fine grained.
Sept. 29	3·9	2	8	Brass 12	0 5	28	Surface covered splendidly and colour very good, withstood very hard buffing, but required only light treatment to finish. Very satisfactory results.
							Electrolyzed solution for 32 consecutive hours with an average current of 10 amperes at 2 volts—90 sq. in. anode surface. Added 24 drops liquid ammonia. Temperature 70 degrees Fahr. Sp. Gr. 1·085. Anode coated with brownish red film when at rest. Salts creep to hooks and remain moist. Solution neutral. Slight sediment at bottom of tank.
Oct. 3	2·8	2	5	Steel 28	24 0	90	Smooth, hard, adherent plate. No evidence of cracks or burning. Thickness approximately $\frac{1}{32}$ inch.
Oct. 3	3·7	2·25	5	Steel Tube 18	2 0	56	Smooth, white, hard plate. Drew the tube from 1″ diameter to $\frac{5}{8}$″ with no damage to deposit.
Oct. 3	4·0	2·25	5	Brass 24	0 10	56	Equal portions were nickel and cobalt plated for 1 hour. The two coatings were then buffed with same stroke. Repeated trials cut the nickel through before the cobalt. In no case did the cobalt expose the brass. Very convincing test.
Oct. 8	3·7	2	5	Brass 12	0 15	28	Plate of good colour, buffed easily to splendid finish; bending, hammering, twisting did not crack or loosen plate.
Oct. 8	4·2	3·5	8	Sheet Steel 72	1 0	90	The cathode consisted of six pieces steel hooked to wire frame. Plate good colour, smooth, hard; did not crack when bent or twisted. Buffed to good finish.
Oct. 14	4·5	4	8	Steel 72	1 0	90	Deposit began with current density of 30 amp. per sq. ft. and gradually increased to 42. After 10 minute run with this current density, plate began to show signs of darkening. This however, was not considered serious until expiration of 20 minutes. Entire surface finished easily on soft buff. While the plate was not a good specimen from a commercial view point, the remarkable efficiency of the bath was clearly demonstrated.

Date	Current density in amps. per sq. dec.	Electro-motive force in volts.	Elect-rode distance in inches.	Cathode area in sq. inches.	Dur-ation of plating.	Anode area in sq. inches.	Remarks.
					hr. min.		
Oct. 15	5·2	4·5	8	Steel 72	1 0	90	Same cathode as last test, arranged differently in frame. Plate grey but not burned; adherent, hard, smooth; did not break when bent; buffed easily to good finish.
Oct. 16	5·2	4·5	8	Cast iron 12	1 0	14	This cathode was a piece of stove casting. Plate on one side only, and of splendid colour over entire surface, the background being equally as white as average nickel plate. When buffed the piece could not be detected from nickel plated piece except by our acquaintance with the fact.
Oct. 31							Current densities as high as 6·3 amperes per square decimetre were employed in several tests, but the results were not such as to merit recording. The solution at this date was proving very efficient in every detail.

Note.—To transform amperes per square decimetre to amperes per square foot, multiply by 9·3.

Ageing Test, Solution I B.

	Grams cobalt per 100 c.c. solution.	Acidity of solution
Solution at end of above series of experiments, Aug. 26, 1914, after 80 hours plating as shown....................................	2·54	Neutral
Same after 95 hours plating.................	2·58	"
" " 110 " "	2·52	"
" " 125 " "	2·55	"
Solution slightly diluted to original volume to make up for evaporation..................		
Same after 140 hours plating.................	2·58	"
" " 155 " "	2·60	"
" " 170 " "	2·62	"

The last 90 hours plating with this solution were performed at a current density of 3·0 to 3·5 amperes per square decimetre. The cathode area was 34·0 square centimetres.

This solution is neutral to litmus paper, but very slightly alkaline as shown by titration with $\frac{N}{10}$ HCl and litmus indicator. The solution remains absolutely constant and the cobalt anode dissolves satisfactorily.

Current Efficiency, Test Solution I B.

Run I.

Current density approx......................	1·0 amp. per sq. dec.
Cathode—brass asphaltum on back—area......	29·2 sq. cms.
Time of run..............................	1 hr. 0 min.
Average current through bath...............	0·300 amp.
Theoretical weight of cobalt deposited.........	0·331 grams.
Weight of cathode before plating.............	55·213 "
" " " after "	55·536 "
Cobalt deposited............................	0·323 "

$$\text{CURRENT EFFICIENCY} \quad \frac{0·323}{0·331} \quad = \quad 97·7\%$$

Run II.

Current density approx......................	1·0 amp. per sq. dec.
Cathode—brass polished both sides..........	47·4 sq. cms.
Time of run...............................	1 hr. 0 min.
Average current through bath...............	·475 amp.
Theoretical weight of cobalt deposited........	0·523 grams.
Weight of cathode before plating.............	54·7641 "
" " " after "	55·2774 "
Cobalt deposited...........................	0·5133 "

$$\text{CURRENT EFFICIENCY} \quad \frac{0·513}{0·523} \quad = \quad 98·2\%$$

Run III.

Current density approx......................	3·0 amp. per sq. dec.
Cathode—brass—area......................	29·5 sq. cms.
Time of run...............................	1 hr. 0 min.
Average current through bath...............	0·907 amp.
Theoretical weight of cobalt deposited........	0·996 grams.
Weight of cathode before plating.............	69·279 "
" " " after "	70·190 "
Cobalt deposited...........................	0·911 "

$$\text{CURRENT EFFICIENCY} \quad \frac{0·911}{0·996} \quad = \quad 91·5\%$$

Run IV.

Current density approx......................	3·0 amp. per sq. dec.
Cathode—brass polished on both sides........	20·0 sq. cms.
Time of run...............................	1 hr. 0 min.
Average current through bath...............	0·599 amp.
Theoretical weight of cobalt deposited........	0·659 grams.
Weight of cathode before plating.............	19·3182 "
" " " after "	19·9234 "
Cobalt deposited...........................	0·5952 "

$$\text{CURRENT EFFICIENCY} \quad \frac{0·595}{0·659} \quad = \quad 90·3\%$$

Run V.

Current density approx......................	3·0 amp. per sq. dec.
Cathode—brass polished on both sides—area..	20·0 sq. cms.
Time of run...............................	1 hr. 0 min.
Average current through bath...............	0·598 amp. per sq. dec.
Theoretical weight of cobalt deposited........	0·659 grams.
Weight of cathode before plating.............	20·2194 "
" " " after "	20·8100 "
Cobalt deposited...........................	0·5906 "

$$\text{CURRENT EFFICIENCY} \quad \frac{0·591}{0·659} \quad = \quad 89·7\%$$

The data of the plating experiments for solution I B show this solution to be remarkable for the extremely high current densities at which satisfactory plates may be obtained. We find no record of nickel plating being accomplished, except under special conditions such as with rotating cathodes, at anything like the same speed. Commercial tests (see p. 65) show that even higher speeds are possible with this solution under the condition of plating practice.

It is fairly well recognized that any improvement in the chemical composition of solutions for nickel plating, in order that a faster rate of deposition may be brought about, would have to be based upon a higher concentration of the nickel ion in the solution.

The inventors of Prometheus, Persels, and other salts for concentrated nickel solutions, no doubt had this in mind. These new baths are of comparatively recent invention, and there is considerable diversity of opinion as to their merits.

The practical plater knows that he can carry in his plating bath 12 oz. of double nickel salt per gallon of water in the summer, and about 9 oz. of the same salt per gallon in the winter, without danger of frequent crystallization. Taking the higher of these figures, we have a bath equivalent to about 80 grams (NH_4) $SO_4 \cdot NiSO_4 \cdot 6H_2O$ per litre. This solution contains approximately 1·5 per cent metallic nickel. On the other hand, using the Prometheus salts as bought on the market, as much as 2 lbs. may be dissolved to the gallon of water without danger of crystallization in the summer. Since this Prometheus salt contains about 28 per cent of $NiSO_4$, this bath will contain approximately 2·6 per cent of metallic nickel in solution. This tremendous increased metal content of the latter bath accounts for the greater speed at which plating is possible with it.

Comparing the cobalt solution I B with this, we note that it contains 200 grams of $CoSO_4 \cdot (NH_4)_2SO_4 \cdot 6H_2O$ to the litre, so that its concentration in metallic cobalt is approximately 3·0 per cent. We would, therefore, expect this solution to be a very rapid plating one, faster than the other as we find it to be, and it has the advantage of being free from magnesium sulphate, boric acid and the like, which are very considerable and necessary constituents of all the concentrated fast plating nickel solutions. Moreover, we find with all solutions of this type that the cobalt bath is a much more rapid plating one than the nickel bath, taking them at the same concentration. Experiments conducted under conditions of present plating practice, demonstrated that this solution I B was capable of plating cobalt satisfactorily at several times the speed that the Prometheus salt was capable of plating nickel. (See p. 65). This comparison was for the best condition for each that was known to the practical plater in charge of the plating establishment in question, and to us as a result of all of our experiments.

SOLUTION I C.

This solution contains cobalt sulphate and ammonium sulphate in molecular proportions, and is intermediate in concentration between I A and I B.

Cobalt sulphate $CoSO_4$ .... 40 grams
Ammonium sulphate $(NH_4)_2SO_4$ .... 34 "
Water....................................1000 c.c.
Total bath........2·5 litres.

Data of Plating Experiments, Solution, I C.

Beginning May 5, 1913, a series of plating experiments were conducted on this bath. The electromotive force across the electrodes was progressively increased so that a series of plates was obtained at current densities from $0 \cdot 30$ to $1 \cdot 25$ amperes per square decimetre in the manner shown in the tables of data for solutions I A and I B. Solution I C is intermediate in concentration between I A and I B, and its properties, as regards the deposition of its metallic content, we found to be correspondingly intermediate between these two. It is not nearly as rapid nor as satisfactory as solution I B.

CONCLUSIONS.

(1) Cobalt plates from these cobalt ammonium sulphate solutions, on brass and iron, are firm, adherent, hard and uniform, and may readily be buffed to a satisfactorily finished surface. They take a very high polish, with a beautiful lustre, which although brilliantly white, possesses a slightly bluish cast.

(2) The specific electrical conductivity of these cobalt ammonium sulphate solutions is very much higher than that of the corresponding nickel solutions.

(3) All of these cobalt plates within the current density ranges described as satisfactory are as smooth, adhesive and generally satisfactory as the best nickel plates.

(4) Solution I A does not lend itself to extremely fast plating like I B, but satisfactory plates may be obtained with it at current densities up to $0 \cdot 80$ amperes per square decimetre.

(5) Solution I A may be used at higher current densities than the corresponding nickel solution, for which Langbein recommends a current density of $0 \cdot 30$ amperes per square decimetre.

(6) Solution I A does not change appreciably in cobalt content or in acidity when used over long periods of time at the recommended current densities.

(7) Solution I B, which is a nearly saturated solution of $CoSO_4 \cdot (NH_4)_2SO_4$, containing 200 grams of $CoSO_4 \cdot (NH_4)_2SO_4 \cdot 6H_2O$ to the litre of water, yields satisfactory cobalt deposits at all current densities up to 4 amperes per square decimetre which is $37 \cdot 2$ amperes per square foot. This very rapid plating was performed in a manner similar to that of common plating practice.

(8) There is no nickel bath operating in the manner of the usual commercial plating procedure at anything like as high a current density as cobalt solution I B. More specifically, the allowable current density with which an adherent firm, smooth, white, hard plate may be obtained with solution I B, without sign of pitting or peeling, and yet which may be readily and satisfactorily finished, is four times that for which the same results may be obtained with the fastest commercial nickel solutions.

(9) Both solutions I A and I B may be used for plating on the usual surface, including brass, iron and steel. No preliminary coating of copper is necessary when plating with these solutions on iron and steel.

(10) Solutions I A and I B may both be used with a large proportion of rolled anodes without becoming acid or depleted in metal.

(11) Solution I B does not change appreciably in cobalt content or in acidity when used over long periods of time at the recommended high current density.

3

(12) The current efficiency of solution I B is extremely high at a current density of 1 ampere per square decimetre. The mean of our measurements which agree very well among themselves, gave a value of 98·0 per cent. The current efficiency of solution I B is as high at 3 amperes per square decimetre as is common for the best nickel solutions that are used in nickel plating practice at very much lower current densities. The average of three current efficiency measurements with solution I B, at 3 amperes per square decimetre, which measurements agreed very well among themselves, was 90·5 per cent.

(13) Solution I C is intermediate in concentration between I A and I B and its properties, as regards speed and quality of the plates to be obtained therefrom, are correspondingly intermediate. It is not nearly so rapid or as satisfactory at high current densities as I B.

(14) Solution I B, when operated slightly alkaline yields plates which are greyish in colour, which peel, pit and show blisters. This solution, when operated acid yields plates, which while fairly adherent, firm and smooth, are dark and freakish.

This bath should be run neutral, for these plates are adherent, firm, smooth, white, hard, yet easily buffed to an excellent finish.

(15) Solution I B requires very little, if any, ageing to put it in condition, but yields satisfactory plates almost from the start.

(16) The "throwing" power of solution I B is remarkably satisfactory.

(17) The anodes in solution I B are remarkably free from a coating such as characterizes nickel anodes.

Solution I B shows so many superior qualities that it seems highly worth while to develop it further, and particularly to study it under exact commercial conditions. This work is reported in a later paragraph under "Commercial Tests with Solution I B," see page 63 et. seq.

SERIES 2: COBALT AMMONIUM SULPHATE SOLUTIONS WITH AN EXCESS OF AMMONIUM SULPHATE.

SOLUTION II.

This solution was prepared with the view to increasing the conductivity by addition of a relatively larger amount of ammonium sulphate. This bath differs from Series 1 in that the $CoSO_4$ and the $(NH_4)_2SO_4$ are not in molecular proportions.

Solution II contained:—

Cobalt sulphate $CoSO_4$...................... 16·7 grams.
Ammonium sulphate $(NH_4)_2SO_4$.............. 56·7 grams.
Water.................................1000 c.c.

If upon preparing this bath it was somewhat too acid, it was neutralized with ammonia. This solution was boiled in preparation and prior to using.

This solution is not nearly saturated in cobalt ammonium sulphate.

Data of Plating Experiments, Solution II.

Cast Anodes.

Date of experiment	Current density in amperes per sq. decimetre.	Electromotive force in volts.	Electrode distance in cms.	Cathode area in sq. cms.	Duration of plating		Character of plate and remarks
1914					hr.	min.	
Sept. 18	0·31	0·91	10	Brass 39·4	1	15	Good white even smooth deposit, requiring almost no buffing to give fine lustre.
Sept. 18	0·56	1·18	10	Brass 33·6	1	15	Good white uniform deposit with fine lustre when polished.
Sept. 18	0·60	1·25	10	33·6	0	45	Same
Sept. 18	0·75	1·35	10	Brass 38·7	0	45	Same
Sept. 24	0·90	1·27	10	38·7	4	0	A bright evenly deposited plate, not lustrous.
Sept. 25	1·0	1·30	10	38·7	3	0	Plate smooth and bright, somewhat spotted.
Sept. 27	1·0	1·55	10	38·7	3	0	This run immediately followed one at current density 1·4. After the former the solution was found rather too acid and was neutralized with ammonia and slightly acidified with boric acid. A dark plate bright with buffing.
Sept. 29	1·0	2·10	20	Brass 38·7	2	0	Plate dark but polished up brightly.
June 29	1·0	2·10	20	27·3	2	0	Deposit dark but buffed up to satisfactory brilliant surface.
July 2	1·0	1·4	10	27·3	2	0	Plate badly burned over entire surface.
July 3	1·0	1·95	20	27·3	2	0	Plate was very black with rough grainy surface when removed from solution. Burned at sides and at bottom.
July 13	1·0	2·05	15	39·3	2	0	A very poor plate black and polished with difficulty.
July 14	1·0	2·05	15		1	30	Plate very unsatisfactory, black and grainy.
July 15	1·0	1·42	15		1	30	Bath stirred by bubbling air through it. Plate not hard, and burned on sides, generally unsatisfactory.
June 26	1·4	1·95	10	27·3	3	0	Plate dark and badly burned.

Solution II does not give good results above 0·90 amperes per square decimetre.

The solution used for the above series of plating experiments was found to be unsatisfactory until boiled. With an unboiled solution, very dark unsatisfactory plates were obtained at current densities between 0·30 and 0·62 amperes per square decimetre.

This solution was not thought of sufficient importance to warrant making a series of tests with rolled anodes, or ageing and efficiency tests.

CONCLUSIONS.

(1) Cobalt plates from this cobalt sulphate and ammonium sulphate solution, with an excess of ammonium sulphate, on brass and iron, are firm, adherent, hard and uniform, and may readily be buffed to a satisfactorily finished surface, within the narrow range recommended for them. They take a very high polish, with a beautiful lustre, which although brilliantly white, possesses a slightly bluish cast.

(2) The specific electrical conductivity of solution II is considerably higher than that of the corresponding nickel solution.

(3) Solution II is not a fast plating solution, and can only be used at current densities up to about 0·90 amperes per square decimetre. This bath is not nearly as rapid nor as satisfactory as others described.

(4)   Solution II is an analogue of one proposed by Langbein for nickelling, of which he says that the nickel deposit piles up, especially in the lower portion of the object.   That is, the lower part of the cathode becomes dull, burned or over-nickelled.   This takes place with the nickel solution at current densities about 0.35 amperes per square decimetre, and consequently the cobalt solution is a very great improvement, as regards speed on the corresponding nickel solution.

(5)   Solution II requires to be boiled at the outset to yield satisfactory plates.   Otherwise the plates are dark, even at low current densities.

(6)   Solution II after operating a number of hours tends to become acid.   This acid may be neutralized with ammonia, and the solution reacidified with boric acid, to yield satisfactory plates.   However, on this account and for others mentioned, this solution is not nearly as satisfactory as some others described.   It is not self sustaining.

SERIES 3: COBALT AMMONIUM SULPHATE WITH AN EXCESS OF AMMONIUM SULPHATE, AND CITRIC ACID.

SOLUTION III A.

A bath which was formerly in extended use for nickel plating is prepared by boiling 34.5 grams nickel sulphate, with 50.3 grams ammonium sulphate, and adding 4.2 grams citric acid, to the litre of water.   Analogous to this solution IIIA was made up containing:—

Cobalt sulphate $CoSO_4$	34.5 grams.
Ammonium sulphate, $(NH_4)_2 SO_4$	50.3 "
Citric acid	4.2 "
Water	1000 c. c.

When solution III A was not found to be very satisfactory, 1.7 grams sodium sulphite were added to the litre.   This latter was solution IIIB. The total amount of the bath prepared was 3 litres.

No reagent was added during the experiments with these solutions either to neutralize them or for any other purpose.

Data of Plating Experiments, Solution III A.

				Cast Anodes.		
Date of experiment.	Current density in amperes per sq. decimetre.	Electromotive force in volts.	Electrode distance in cms.	Cathode area in sq. cms.	Duration of plating.	Character of plate and remarks.
				Brass	hr. min.	
June 10	0.20	1.5	10	27.3	1 30	Uniform deposit but very dark, which was susceptible to bright polish with buffing.
June 25	0.50	1.25	10	27.3	3 0	Plate black but smooth.
June 29	0.50	1.39	20	27.3	1 0	Plate was slightly dark on removing from bath, and buffed to a beautiful bright, clean surface.
June 30	0.80	2.13	20	27.3	1 30	Plate scaley, peeling easily, not satisfactory.
June 25	1.0	2.35	10	27.3	1 30	Plate uniformly black and unsatisfactory.

This bath was not thought to be of sufficient importance to warrant running a series of experiments with rolled anodes.

SOLUTION III B.

Cobalt sulphate $CoSO_4$......................	34·5	grams.
Ammonium sulphate $(NH_4)_2 SO_4$..............	50·3	"
Citric acid	4·2	"
Sodium sulphite Na_2SO_3	1·7	"
Water......................................	1000 c. c.	
Total bath...........................	3 litres.	

Data of Plating Experiments, Solution III B.

Date of experiment.	Current density in amperes per sq. decimetre.	Electromotive force in volts.	Electrode distance in cms.	Cathode area in sq. cms.	Duration of plating.		Character of plate and remarks.
				Cast Anodes.			
July 20	0·30	1·06	10	Steel 60·0	1	15	Plate satisfactory and adherent.
July 10	0·30	0·65	15	Brass 27·3	2	0	Very smooth, hard plate which buffed to bright silvery white finish.
July 20	0·40	1·0	10	Steel 60·0	1	15	Satisfactory smooth, adherent plate.
July 27	0·40	1·0	10	Brass 31·5	2	30	Smooth, uniform plate, slightly dark in colour which peeled at edges after removal from bath.
June 30	0·50	1·25	20	Brass 27·3	1	30	Bright, hard plate, good colour after being polished. No signs of pitting.
July 3	0·50	0·86		Brass 27·3	1	30	Extremely bright smooth plate, free from all flaws which buffed to beautiful surface.
Aug. 3	0·80	0·96	10	Brass 39·3	2	0	A very white uniform and satisfactory plate except for slight scaling on one side.
July 11	0·80	1·35	15	Brass 27·3	1	30	Very smooth, even plate, which buffed to satisfactory surface.
July 13	1·0	1·65	15	27·3	1	30	Plate scaled off. Unsatisfactory.
July 2	1·0	2·07	20	27·3	3	0	Clean, hard uniform deposit, which buffed to a beautiful surface.
June 26	1·36	1·86	10	Brass 31·5	3	0	Smooth, uniform plate but dark in colour.
June 26	1·7	2·1	10	24	1	30	Plate unsatisfactory. Split.
				Rolled Anodes.			
June 17	1·0	2·5	15	Brass 37	1	0	Clean smooth plate, but black in appearance.

Ageing Test, Solution III B.

	Grams cobalt per 100 c.c. solution	Acidity of solution
Solution at end of above series of experiments, Aug. 26, 1914, after 24 hours plating as shown..............................	1·11	Very slightly alkaline.
Same after 39 hours plating.................	1·10	Same
" " 54 " "	1·14	Very slightly alkaline.
" " 69 " "	1·10	Same
Diluted slightly to make up for loss of water by evaporation............................	1·11	"
Same after 84 hours plating.................	1·12	"
" " 99 " "	1·13	"
" " 114 " "	1·14	"

The last 90 hours plating was at a current density of $1\cdot0$ ampere per square decimetre. The cathode area was $34\cdot0$ square centimetres. The cobalt content and alkalinity of this solution remained approximately constant.

Current Efficiency Test, Solution III B.

Run I.

Current density..............................	$1\cdot0$ amp. per sq. dec.
Cathode—brass—asphaltum on back-area	25 sq. cms.
Time of run................................	30 min.
Average current through bath...............	$0\cdot250$ amp.
Theoretical weight of cobalt deposited........	$0\cdot138$ grams.
Weight of cathode before plating.............	$65\cdot7852$ "
" " " after "	$65\cdot9192$ "
Cobalt deposited............................	$0\cdot1340$ "

$$\text{CURRENT EFFICIENCY} \quad \frac{0\cdot134}{0\cdot138} = 97\cdot3 \ \%$$

Run II.

Current density..............................	$1\cdot0$ amp. per sq. dec.
Cathode—brass polished both sides—area....	$32\cdot6$ cms.
Time of run................................	1 hr. 0 min.
Average current through bath...............	$0\cdot328$ amp.
Theoretical weight of cobalt deposited........	$0\cdot361$ grams.
Weight of cathode before plating.............	$24\cdot6807$ "
" " " after "	$25\cdot0370$ "
Cobalt deposited............................	$0\cdot3563$

$$\text{CURRENT EFFICIENCY} \quad \frac{0\cdot356}{0\cdot361} = 98\cdot6 \ \%$$

For comparison with solutions IIIA and IIIB, a set of experiments, on the corresponding nickel bath was made, which is solution III C.

SOLUTION III C.

Nickel sulphate $NiSO$.......................	$34\cdot5$ grams.
Ammonium sulphate $(NH_4)_2 SO_4$..............	$50\cdot3$ "
Citric acid.................................	$4\cdot2$ "
Water.....................................	1000 c.c.
Total bath 3 litres.	

Data of Plating Experiments Solution III C.

Cast Anodes.

Date of experiment.	Current density in amperes per sq. decimetre.	Electro-motive force in volts.	Electrode distance in cms.	Cathode area in sq. cms.	Duration of plating.		Character of plate and remarks.
1914 July 16,	0·30	1·25	10	Steel 39·3	1	30	Deposit scaled easily.
July 13	0·34	1·28	10	Brass 39·3	1	30	Plate white and uniform.
July 31	0·50	1·45	10	Brass 39·4	2	0	Plate satisfactory except for pitting, and easily polished. It is difficult to get a free plate from hydrogen pits from this bath except at very low current densities.
July 22	0·50	1·45	10	39·3	1	30	Brilliant white plate as taken from bath, which buffed satisfactorily. This plate was decidedly softer than one from Co bath III E under same date, which was run and buffed simultaneously with it.
July 22	0·50	1·45	10	39·3	1	30	Same
July 23	0·90	2·35	10	46·6	1	0	Plate bright metallic lustre as taken from solution. Readily buffed to satisfactory mirror surface. Whiter and softer than Co plate solution III run simultaneously.
July 14	1·0	2·15	10	Brass 39·3	1	30	Plate bright but pitted.
July 14	1·0	2·6	10	39·3	1	30	Plate bright but pitted on surface yielding a beautiful surface after buffing.
July 14	1·0	2·6	10	Brass 37·1	1	30	Satisfactory plate except for few small pits.
July 24	1·0	2·36	10	Brass 39·3	1	30	Plate showed darkness on edges, although it buffed satisfactorily.
July 24	1·0	2·36	10	Brass 39·3	1	30	Plate white as removed from bath but difficult to polish to satisfactory mirror surface, slightly pitted.
July 25	1·07	2·56	10	Brass 39·3	1	30	Plate scaled slightly at lower edge.
July 28	1·25	2·60	10	Brass 39·4	1	45	Plate dull in colour and buffed to satisfactory mirror surface with difficulty.
July 29	1·50	2·75	10	Brass 27·3	2	0	Plate dull and burned.
July 14	1·5	2·75	10	Brass 27·3	1	30	Plate bright with pits and scaling at bottom.

Rolled Anodes.

Date of experiment.	Current density in amperes per sq. decimetre.	Electro-motive force in volts.	Electrode distance in cms.	Cathode area in sq. cms.	Duration of plating.		Character of plate and remarks.
July 31	0·50	2·85	10	Brass 39·3	2	0	Very satisfactory plate white as removed from solution, and readily polished to mirror surface.
July 31	0·80	3·30	10	Brass 39·3	2	0	Same

Current Efficiency Test, Solution III C.

Run I.

Current density approx.................... 1·0 amp. per sq. dec.
Cathode—brass polished on both sides—area... 24·2 sq. cms.
Time of run.................... 1 hr. 0 min.
Average current through bath................ 0·242 amp.
Theoretical weight of cobalt deposited........ 0·266 grams.
Weight of cathode before plating............ 21·1806 "
 " "   " after " .... 21·4279 "

Cobalt deposited........................ 0·2473

$$\text{CURRENT EFFICIENCY} \quad \frac{0·247}{0·266} = 92·8\%$$

Run 11.

Current density approx............................ 1·0 amp. per sq. dec.
Cathode—brass—area........................... 40·0 sq. cms.
Time of run..................................... 1 hr. 0 min.
Average current through bath.................. 0·400 amp.
Theoretical weight of cobalt deposited.......... 0·440 grams.
Weight of cathode before plating.............. 34·3341 "
"   "   after   "   .... 34·7404 "

Cobalt deposited............................... 0·4063 "

$$\text{CURRENT EFFICIENCY} \quad \frac{0\cdot406}{0\cdot440} \; = \; 92\cdot2\%$$

SOLUTION III D.

Cobalt sulphate $CoSO_4$ ....25·7 grams.
Ammonium sulphate....................... 31·5 "
Citric acid............................... 4·5 "
Water.................................... 1000 c.c.

This bath resembles solution III A, but is considerably less concentrated in cobalt sulphate and ammonium sulphate.

There were no reagents added to this solution after it was made up for neutralizing or any other purpose.

Data of Plating Experiments, Solution III D

						Cast Anodes.
Date of experiment.	Current density in amperes per sq. decimeter.	Electromotive force in volts.	Electrode distance in cms.	Cathode area in sq. cms.	Duration of plating.	Character of plate and remarks.
1914 July 14	0·30	1·06	15	Brass 37	3	0	Black as removed from solution, but buffed satisfactorily, except for small pin holes.
July 7	0·50	1·36	15	37	3	0	Plate dark as removed from solution, but buffed satisfactorily.
July 7	0·50	1·33	15	37	3	0	Same
July 7	0·70	1·80	15	37	3	0	Plate bright as removed from bath, but slightly rough.
July 11	0·90	1·89	15	37	3	0	Plate dull and pitted. Current density too great.
July 11	0·90	1·90	15	37	3	0	Deposit dull and pitted. Current density too great. In polishing this plate the extreme hardness of the Co deposit as compared with Ni was particularly noticeable. The same result was noticed to marked extent throughout these plating experiments.
July 17	1·0		15	Brass 37	1	0	A clean smooth plate but dark.
July 7	1·0	2·02	15	37	3	0	Current density seemed too high as plate was very dull.
July 7	1·0	2·0	15	37	3	0	Plate dull and pitted. Current density too great.
July 13	1·0	2·0	15	37	3	0	Deposit dull and pitted. Current density too great. In polishing this plate the extreme hardness of the Co deposit as compared with Ni was particularly noticeable. The same result was noticed to marked extent throughout these plating experiments.

This bath has been used for plating approximately 35 hours, after which it becomes somewhat alkaline. In comparison, solution III C, which is the corresponding nickel bath, was tested, after running with the same current for an identical time. It was found to be practically in the same condition as regards alkalinity as at the start. Cast anodes were used in both of these baths throughout these runs, which were made partly for the purpose of testing the relative solubility of the anodes. This result confirmed a conclusion to be generally drawn from all our experiments, that in baths of this type the cast cobalt anodes are more soluble than the cast nickel anodes.

SOLUTION III E.

Cobalt sulphate	$CoSO_4$	78·5 grams.
Ammonium sulphate	$(NH_4)_2SO_4$	129·7 "
Citric acid		13·7 "
Water		1000 c.c.

This bath is more than twice as saturated in cobalt sulphate as III A, and more than three times as saturated in cobalt sulphate as III D.

Data of Plating Experiments, Solution III E.

Date of experiment.	Current density in amperes per sq. decimetre.	Electromotive force in volts.	Electrode distance in cms.	Cathode area in sq. cms.	Duration of plating.		Character of plate and remarks.
1914 July 21,	0·50	0·76	10	Brass 39·3	2	0	Very bright uniform plate.
July 22	0·50		10	39·3	2	0	Plate bright and satisfactory. Considerably harder than Ni plate run at same time, see III C, July 22, 0·50 amp.
July 28	0·62	0·87	10	Steel 20	3	0	Plate dark grey with metallic lustre as removed from solution. Required very little buffing to produce mirror surface. Plate satisfactory in all respects.
July 22	0·70	0·90	10	Brass 39·3	2	0	Plate bright and satisfactory. Considerably harder than Ni plate run at same time, see III C July 22, 0·70 amp.
July 21	0·80	0·82	10	Brass 39·3	2	0	Very smooth though somewhat dark plate as removed from bath. Polished to brilliant mirror surface.
July 23	0·80	0·97	10	39·3	2	0	Same, see III C, July 23, 0·80 amps.
July 23	0·90	1·0	10	Brass 46·6	1	0	A satisfactory plate, readily buffed to mirror surface. Not as white when removed from bath as Ni plate run simultaneously, see III C, July 23, 0·90 amperes Co plate much harder than corresponding Ni plate.
July 21	1·0	1·03	10	39·3	2	0	Very smooth though somewhat dark plate as removed from bath. Polished to brilliant mirror surface.
July 24	1·0	1·05	10	39·3	2	0	Plate scaled off when put in boiling water after removed from bath.
July 24	1·0	1·05	10	39·3	2	0	Plate dark though lustrous with no sign of scaling as removed from bath. This plate was darker than corresponding Ni plate as removed from the bath. See III C, July 24, 1·0 amp., although the Ni plate required more polishing to produce a mirror surface.
July 25	1·07	1·19	10	39·3	2	0	Plate rather dark and showed signs of burning.
July 28	1·25	1·31	10	Brass 39·3	1	45	Plate dark grey, but required very little buffing to produce mirror surface.
July 29	1·5	1·35	10	27·3	2	0	Plate somewhat dark as taken from solution, but readily buffed to perfect mirror surface. Co plate considerably harder than corresponding Ni plate when both were ground.
July 31	...	....	..	....	..	..	This bath becoming alkaline, which was neutralized by small addition of citric acid.

Ageing Test, Solution III E.

	Grams cobalt per 100 c.c. solution.	Acidity of solution.
Solution at end of above series of experiments, Aug. 26, 1914, after 26 hours plating as shown........	1·72	Neutral.
Same after 41 hours plating	1·81	Very slightly alkaline.
" 56 "	1·95	Slightly alkaline.
" 71 "	2·12	Neutral.
After 86 hours plating a slight amount of water added to make up for loss due to evaporation.............	1·95	Slightly alkaline.
Same after 101 hours plating	1·99	Slight increase in alkalinity.
" 116 "	2·11	Further increase in alkalinity.
" 131 "	2·17	Still further increase in alkalinity.

The last 45 hours of plating was at a current density of 3·0 amperes per square decimetre.   The 45 hours preceding this was at a current density of 4·75 amperes per square decimetre.

This solution is gradually becoming more alkaline and its cobalt content is increasing.   In its present state, this solution is obviously changing too rapidly to be satisfactory, but with an increased number of rolled anodes replacing cast ones, it might be used.

Current Efficiency Test, Solution III E.

Run I.

Current density approx......................	0·80 amp. per sq. dec.
Cathode—brass polished both sides—area.....	40·0 sq. cms.
Time of run.................................	45 min.
Average current through bath................	0·321 amp.
Theoretical weight of cobalt deposited........	0·265 grams.
Weight of cathode before plating.............	35·8816 "
" " after "	36·1435 "
Cobalt deposited...........................	0·2619 "

$$\text{CURRENT EFFICIENCY} \quad \frac{0·262}{0·265} = 98·8 \%$$

Run II.

Current density approx......................	0·80 amp. per sq. dec.
Cathode—brass polished both sides—area......	40·0 sq. cms.
Time of run.................................	1 hr. 0 min.
Average current through bath................	0·321 amp.
Theoretical weight of cobalt deposited........	0·353 grams.
Weight of cathode before plating.............	36·1435 "
" " after "	36·4949 "
Cobalt deposited...........................	0·3514

$$\text{CURRENT EFFICIENCY} \quad \frac{0·351}{0·353} = 99·4 \%$$

31

CONCLUSIONS.

(1) Cobalt plates from these cobalt-ammonium sulphate solutions with an excess of ammonium sulphate, and in the presence of citric acid, on brass and iron, are firm, adherent, hard and uniform and may readily be buffed to a satisfactory finish, within the current density range recommended. They take a very high polish, with a beautiful lustre, which, although brilliantly white, possesses a slightly bluish cast.

(2) Solutions III A, III B, and III D do not lend themselves to extremely fast plating, but satisfactory plates may be obtained from them at current densities up to 0·80, 1·0 and 0·80 amperes per square decimetre respectively.

(3) Solution III B maintained itself substantially constant as regards cobalt content and alkalinity during 114 hours of plating. The current efficiency of solution III B is satisfactorily high.

(4) Solution III E, which is very much more concentrated in cobalt than the other solutions of this series, yields satisfactory plates at all current densities up to 1·5 amperes per square decimetre.

(5) Solution III E, when used with cast anodes, gradually becomes more concentrated in metallic cobalt content, and increasingly alkaline. For this reason this solution can only be used with care, and is probably not satisfactory for general commercial plating purposes.

(6) The current efficiency of solution III E is extraordinarily high as compared with that of the usual commercial nickel plating solutions.

(7) The current efficiencies of all the solutions of Series 3 are high, and well over 90 per cent at the recommended current densities.

(8) A number of experiments with solution III C and solution III E were run simultaneously, with the same current density, electrode distance and electrode area. The electromotive force across the nickel bath was from two to two and a half times as great as that across the cobalt bath. This is evidence, as noted throughout these experiments, of the greater conductivity of the cobalt solutions.

(9) Solution III C is very concentrated in nickel salt, and is more satisfactory than the corresponding cobalt solution of the same concentration. However, the corresponding cobalt solution is not nearly saturated. The comparison with an equally saturated similar cobalt solution III E, is in favour of the cobalt solution, for it will operate at higher current densities than III C, and yields a plate which is just as satisfactory in appearance and harder.

(10) These solutions operated alike plating on iron, steel and brass, for which metals only the above conclusions apply.

(11) Cobalt anodes are more readily soluble than nickel anodes in the solutions of Series 3, and consequently a larger proportion of rolled cobalt anodes may be used than is the case with nickel.

SERIES 4: COBALT AMMONIUM SULPHATE AND AMMONIUM CHLORIDE SOLUTIONS.

The baths of Series 4 were prepared in a manner similar to that described under Series 1.

It is stated in the literature[1] that baths containing chlorides or nitrates are not suitable for nickelling over iron. They are, however, well adapted to the rapid light nickelling of cheap brass articles. Solution IV

---

[1] 3a Langbein Electro-deposition of Metals, VI Edition, Revised, p. 254.

is the cobalt analogue of one sometimes used in nickelling practice for work of this lighter kind.

SOLUTION IV A.

Cobalt sulphate $CoSO_4$...................23·5 grams
Ammonium sulphate $(NH_4)_2 SO_4$..........20·0 "
Ammonium chloride $NH_4 Cl$...............30·0 "
Water....................................1000 c.c.
   Total bath........................... 2·5 litres.
   Sp. Gr....... 1·043   = 6·0° Be.

No reagents were added to this solution after it was made up either to change its relation or for other purposes.

Data of Plating Experiments, Solution IV A.

				Cast Anodes.		
Date of experiment.	Current density in amperes per sq. decimetre.	Electromotive force in volts.	Electrode distance in cms.	Cathode area in sq. cms.	Duration of plating.	Character of plate and remarks.
1914 Sept. 21	0·50	0·85	10	Brass 34·6	2 hr. 0 min.	Good, smooth uniform deposit, requiring but little buffing to bring to mirror surface.
June 11	0·55	0·65	10	Brass 27·3	3 0	Same
Sept. 21	0·75	0·94	10	Brass 25·8	1 0	Good white plate, uniform and velvety. Readily buffed to mirror surface.
June 26	1·0	1·1	10	Brass 24·0	1 30	Same
Sept. 21	1·0	1·45	10	Brass 35·5	1 0	Plate uniform, but grey. Good finish when polished, but somewhat gas pitted.
June 25	1·1	1·2	10	Brass 24·0	2 45	Good, smooth, uniform deposit, velvety in appearance, which took a brilliant silvery polish.
June 14	1·1	1·15	10	Brass 27·5	2 0	This run was to determine nature of deposit at twice the current density recommended for similar Ni bath. A good smooth, uniform deposit obtained over entire surface of cathode. Velvety appearance.
June 25	1·65	1·75	10	Brass 31·5	1 30	Deposit even and uniform, but slightly dark in centre, which buffed to a very satisfactory surface.
June 26	1·9	1·92	10	Brass 31·5	3 0	Plate not satisfactory, somewhat burned on edges.

SOLUTION IV B.

Cobalt sulphate $CoSO_4$.................... 23·5 grams.
Ammonium sulphate $(NH_4)_2SO_4$............ 20·0 "
Ammonium chloride $NH4_4Cl$............... 18·2 "
Boric acid................................ 9·5 "
Water....................................1000 c.c.
   Total bath........................... 2·5 litres.

This solution is slightly acid to litmus. Before using, this acidity was neutralized with ammonia, then boric acid added until the solution was slightly acid again.

Data of Plating Experiments, Solution IV B.

Cast Anodes.

Date of experiment.	Current density in amperes per sq. decimetre.	Electromotive force in volts.	Electrode distance in cms.	Cathode area in sq. cms.	Duration of plating.		Character of plate and remarks.
1914 Aug. 4	0·50	0·90	10	Iron 45·4	hr. 0	min. 40	Plate grey, but very smooth and bright as removed from the bath. Readily buffed to satisfactory mirror surface.
June 12	0·55	0·75	10	Brass 23·3	3	0	Plate hard and uniform, and much whiter than solution IV A.
Sept. 21	1·0	1·77	10	Brass 36·0	1	0	Good, uniform white deposit. Satisfactory finish with little buffing.
June 12	1·10		10	Brass 27·3	3	0	Plate hard and uniform, much whiter than IV A.
Sept. 22	1·25	1·67	10	Brass 28·1	1	0	Good plate, slightly dark at edges. No burning at this current density.
June 14	1·5	2·5	12	Brass 30·0	1	0	Plate at first bright, but later blackened along edges, showing signs of burning.
Sept. 22	1·5	2·27	10	Brass 34·8	0	30	Very dark and badly burned at edges.
June 27	1·6	2·22	20	Brass 31·5	2	30	Smooth plate, but not uniform. Brighther nearer bottom than top.
June 26	1·65	1·86	10	Brass 24·0	2	0	Smooth, uniform plate, somewhat whither than plate solution IV A, June 26, Current Density 1 ampere per square decimetre. Plate somewhat dark at edges.
July 29	1·70	2·15	10	Brass 27·3	1	30	Gas at cathode. Plate unsatisfactory, grainy and burned.

Ageing Test, Solution IV B.

Solution at end of above series of experiments, Aug. 26, 1914, 16 hours plating as shown .... 0·76 Neutral

Same after 31 hours plating..................... 0·80 "

Same after 46 hours plating..................... 0·74 "

After 61 hours plating diluted solution to make up for loss due to evaporation................... 0·75 "

Same after 77 hours plating..................... 0·96 "

This solution is neutral to litmus paper and slightly alkaline by titration with $\frac{N}{10}$ HCl and litmus indicator. It remained constant in acidity throughout the runs.

Current Efficiency Test, Solution IV B.

Run I.

Current density approx..................... 1·0 amp. per sq. dec.

Cathode—brass, asphaltum on back—area..... 36·5 sq.cms.

Time of run............................. 1 hr. 0 min.

Average current through bath............... 0·368 amp.

Theoretical weight of cobalt deposited........ 0·405 grams.

Weight of cathode before plating............ 34·7766 "

Weight of cathode after plating.............. 35·1502 "

Cobalt deposited........................ 0·3736

$$\text{CURRENT EFFICIENCY} \ \frac{0·374}{0·405} = 92·3 \ \%$$

Run II.

Current density approx.....................	0·42 amp. per sq. dec.
Cathode—brass, polished both sides—area.....	43·4 sq. cms.
Time of run............................	1 hr. 0 min.
Average current through bath...............	0·182 amp.
Theoretical weight of cobalt deposited........	0·200 grams.
Weight of cathode before plating............	33·7041 "
Weight of cathode after plating.............	33·8936 "

Cobalt deposited........................... 0·1895

$$\text{CURRENT EFFICIENCY} \quad \frac{0·190}{0·200} = 95·0 \ \%$$

Run III.

Current density approx.....................	1·0 amp. per sq. dec.
Cathode—brass polished both sides—area......	51·0 sq. cms.
Time of run............................	1 hr. 0 min.
Average current through bath...............	0·512 amp.
Theoretical weight of cobalt deposited........	0·563 grams.
Weight of cathode before plating............	57·0320 "
Weight of cathode after plating.............	57·5464 "

Cobalt deposited........................... 0·5144

$$\text{CURRENT EFFICIENCY} \quad \frac{0·514}{0·563} = 91·2 \ \%$$

Conclusions.

(1) Solution IV A yields satisfactory cobalt plates on brass and iron at all current densities up to about 1·5 ampere per square decimetre.

(2) The plates of solution IV A buffed to a brilliant surface, similar to that described under conclusion 3 below.

(3) Solution IV B gave satisfactory plates at all current densities up to 1·25 amperes per square decimetre on brass and iron which are firm, adherent, hard and uniform, and which may readily be buffed to a satisfactorily finished surface, they take a very high polish with a beautiful lustre, which, although brilliantly white, possesses a slightly bluish cast.

(4) Solution IV A and IV B are moderately rapid plating baths, but not, however, nearly as rapid as solution I B, and solution XIII B.

(5) Solution IV B is considerably more rapid than the corresponding nickel bath, the latter working best at a current density of about 0·55 amperes per square decimetre.[1]

(6) These solutions operated alike, plating on iron, steel and brass, for which metals only the above conclusions apply.

(7) The cobalt content and the neutrality of solution IV B do not change appreciably with prolonged usage.

(8) The current efficiency of solution IV B is satisfactorily high, the average value of three measurements, agreeing well among themselves, being 92·6 per cent.

(9) Solution IV B, with boric acid, yields somewhat whiter plates than solution IV A, but solution IV A may be operated at a somewhat higher current density.

---

[1] Electrodeposition of Metals, Langbein, 6th Edition, Revised, p. 253.

SERIES 5 : COBALT CHLORIDE AND AMMONIUM CHLORIDE SOLUTIONS.

SOLUTION V.

A bath was prepared analogous to the nickel-ammonium chloride bath which has been largely favoured for nickelling over zinc. This solution is also used for dark nickelling. The salts as given below were dissolved in luke warm water, and a sufficient amount of ammonia added to leave the solution slightly acid or just neutral.

Cobalt chloride $CoCl_2$..................... 54·8 grams.
Ammonium chloride NH_4 Cl............... 54·8 "
Water..................................1000 c.c.
Total bath.......................... 2·5 litres

No reagent was added to this solution to change its reaction after it was made up.

Data of Plating Experiments, Solution V.

Date of experiment.	Current density in amperes per sq. decimetre.	Electromotive force in volts.	Electrode distance in cms.	Cathode area in sq. cms.	Duration of plating.		Character of plate and remarks.
				Cast Anodes.			
1914 June 17	0·30	0·96	10	Brass 27·3	2	20	Plate dark grey in colour.
June 27	0·30	0·66	10	Brass 27·3	2	30	Dark, even plate which buffed up satisfactorily.
June 18	0·37	0·55	10	Brass 27·3	3	20	Deposit dull grey, spotted and rough.
June 18	0·37	0·55	10	Brass 27·3	3	20	Same as above.
June 20	0·50	0·75	10	Brass 27·3	3	20	Plate dark and spotted.
June 22	0·50	0·69	10	Brass 27·3	2	30	Heavy black deposit, which buffed up satisfactorily.
June 29	0·90	1·76	20	Brass 27·3	2	0	Plate very dark and slightly pitted, but buffed to satisfactory surface.
July 15	1·0	1·15	11	Brass 37	1	0	Plate somewhat dull when removed from solution, but buffed to satisfactory surface.
				Rolled Anodes.			
Aug. 3	0·50	0·70	10	Iron 45·4	1	15	Plate very smooth, bright, dark grey in colour as removed from bath, readily buffed to mirror surface.
July 17	0·8	1·0	15	Brass 27·3	1	0	Smooth, bright plate, slightly pitted at bottom.
July 17	1·0	1·1	16·5	Brass 27·3	1	0	Smooth, uniform plate, dark on removal from solution, but bright after buffing.
Aug. 5	1·0	0·95	10	Iron 41·2	2	20	Plate smooth and dull grey, no sign of burning as removed from bath, somewhat pitted. Polished readily to mirror surface.
				Cast Anodes Plating on Zinc.			
Sept. 25 1914	0·50	0·75	10	Cast zinc 26·6	1	30	Bright deposit, almost as if buffed when removed from solution. Mechanical agitation.
Sept. 24	0·50	0·76	10	Same	1	15	Solution stirred mechanically to keep off hydrogen bubbles. Good white, uniform plate, readily buffed to mirror surface.
Sept. 21	0·75	1·06	10	Same	1	0	No agitation, poor plate, crystalline.
Sept. 25	0·75	0·85	10	Same	1	15	Smooth, uniform deposit, looked almost as if buffed when removed from solution.

Continued

Date of experiment.	Current density in amperes per sq. decimetre.	Electro-motive force in volts.	Electrode distance in cms.	Cathode area in sq. cms.	Duration of plating.		Character of plate and remarks.
Sept. 25	0·75	1·0	10	Sheet zinc 28·8	1	0	Very smooth on portion of cathode where agitation removed hydrogen bubbles; rough on part away from agitation where hydrogen clung.
Sept. 23	1·0	0·95	10	Cast zinc 26·6	1	0	Plate kept free of gas bubbles by rubbing for 10 mins. At end of this time plate satisfactory and smooth. At end of hour without further brushing off of bubbles, plate rough and crystalline.
Sept. 24	1·0	0·95	10	Same	1	0	Solution stirred mechanically to keep off hydrogen bubbles, but in spite of this, at this current density, gas pits appeared at top and bottom.
Sept. 24	1·0	0·95	10	Same	1	0	Same
Sept. 24	1·0	0·95	10	Cast zinc 26·6	1	0	Solution stirred mechanically to keep off hydrogen bubbles, but in spite of this, at this current density, gas pits appeared at top and bottom.
Sept. 22	2·0	1·48	10	Same	0	30	Decidedly crystalline deposit, very poor plate. Gas formed and adhered to cathode.

CONCLUSIONS.

(1) Solution V was not found to be satisfactory for obtaining a bright characteristic cobalt plate at any current density up to one ampere per square decimetre, either with rolled or cast anodes, plating on brass and iron. This refers to plating in the normal manner and without agitation of the solution. This solution could be used on brass and iron if dark cobalting were required.

(2) Rolled anodes are required for a satisfactory deposit with the nickel analogue of this solution. They improve the deposit in the case of cobalt, but it is by no means satisfactory, either in speed or quality of deposition, as compared with other solutions, as I B and XIII B.

(3) Solution V may be used satisfactorily to cobalt on zinc, provided there is a sufficient mechanical agitation to remove hydrogen bubbles from the surface of the cathode. With this provision the cobalt plates are firm, adherent, hard and uniform, and of a polished appearance as removed from the solution.

SERIES 6: COBALT AMMONIUM SULPHATE WITH BORIC ACID SOLUTIONS.

Solution VI A is analogous to a solution recommended by Weston for nickel baths. It has, however, never found extended usage in commercial plating for the reason that the nickel solution, after working faultlessly for a comparatively short time, begins to fail, yielding a blackened deposit.

Solution VI B is the nickel solution corresponding to the cobalt solution VI A. After solution VI A was found to be unsatisfactory, its metal content was increased by further addition of cobalt sulphate. This solution was called VI C.

In addition a much more concentrated solution of the same series, VID was prepared and studied.

SOLUTION VI A.

Cobalt sulphate $CoSO_4$ . . . . 14·8 grams.
Ammonium sulphate $(NH_4)_2SO_4$ . . . . 12·6 "
Boric acid . . . . 18·8 "
Water . . . . 1000 c.c.
　　　Total bath . . . . 2·5 litres.

No additions were made to this solution after it was made up, to change its reaction.

Data of Plating Experiments, Solution VI A.

Cast Anodes.

Date of experiment.	Current density in amperes per sq. decimetre.	Electromotive force in volts.	Electrode distance in cms.	Cathode area in sq. cms.	Duration of plating.		Character of plate and remarks.
1914 June 13	0·30	0·96	10	Brass 27·3	2	20	Deposit lustrous but dark and not very smooth.
June 15	0·60	1·70	10	Brass 27·3	2	20	Plate lustrous, but spotted; hydrogen gas given off freely.
June 27	0·63	1·25	10	Brass 31·5	2	30	Plate smooth and fairly lustrous, but dark. Split on edges.
June 30	0·90	3·5	20	Brass 27·3	2	0	Plate unsatisfactory, pitted and burned.
June 30	1·0	3·27	20	Brass 27·3	2	0	This plate badly burned and unsatisfactory. Current density too great.
June 15	1·0	3·6	15	Brass 24·0	1	30	Plate badly burned at the edges.

Rolled Anodes.

July 16	0·5	2·0	15	Brass 37·3	1	30	Bright, smooth and uniform plate.
July 16	0·8	2·3	13	Brass 37·3	2	0	Unsatisfactory plate, burned along edges and peeled.
July 16	0·8	2·5	15	Brass 27·1	1	0	Same as above.

SOLUTION VI B.

Nickel sulphate $NiSO_4$. 14·8 grams.
Ammonium sulphate $(NH_4)_2SO_4$. . . . . 12·6 "
Boric acid. . . . . 18·8 "
Water. . . . . 1000 c.c.
Total bath. . . . . 2·5 litres

No reagent was added to this solution after it was made up, to change its reaction.

Data of Plating Experiments, Solution VI B.

Cast Anodes.

Date of experiment.	Current density in amperes per sq. decimetre.	Electromotive force in volts.	Electrode distance in cms.	Cathode area in sq. cms.	Duration of plating.		Character of plate and remarks.
July 15	0·30	1·55	10	Brass 27·3	2	0	A very white even deposit, so glossy when removed from solution that it required almost no buffing.
July 15	0·50	2·0	10	Brass 27·3	2	0	Very bright even plate, with no trace of burning.
July 29	0·60	2·27	10	Brass 39·3	1	30	Plate bright in centre but signs of burning at edges.
July 29	0·80	2·36	10	Brass 39·3	1	30	Plate bright in centre but burned at edges and corners.
July 29	1·0	3·01	10	Brass 39·3	1	30	Plate very dark and grainy, showing marked burning over entire surface but especially at edges.

Current Efficiency Test, Solution VI B.

Run I.

Current density approx......................	0·50 amp. per sq. dec.
Cathode—brass polished both sides—area......	40·0 sq. cms.
Time of run.................................	45 min.
Average current through bath................	0·200 amp.
Theoretical weight of nickel deposited.........	0·165 grams.
Weight of cathode before plating.............	39·1777 "
Weight of cathode after plating..............	39·3174 "
Nickel deposited...........................	0·1397

$$\text{CURRENT EFFICIENCY} \quad \frac{0·140}{0·165} = 84·8\%$$

Run II.

Current density approx......................	0·50 amp. per sq. dec.
Cathode—brass polished both sides—area......	40·0 sq. cms.
Time of run.................................	1 hr. 0 min.
Average current through bath................	0·201 amp.
Theoretical weight of nickel deposited.........	0·221 grams.
Weight of cathode before plating.............	39·3174 "
Weight of cathode after plating..............	39·5307 "
Nickel deposited...........................	0·2133 "

$$\text{CURRENT EFFICIENCY} \quad \frac{0·213}{0·221} = 96·4\%$$

Run III.

Current density approx......................	0·50 amp. per sq. dec.
Cathode—brass polished both sides—area......	63·5 sq. cms.
Time of run.................................	1 hr. 3 min.
Average current through bath................	0·365 amp.
Theoretical weight of nickel deposited.........	0·420 grams.
Weight of cathode before plating.............	52·8654 "
Weight of cathode after plating..............	53·2356 "
Nickel deposited...........................	0·3702 "

$$\text{CURRENT EFFICIENCY} \quad \frac{0·370}{0·420} = 87·8\%$$

Run IV.

Current density approx......................	0·50 amp. per sq. dec.
Cathode—brass polished both sides—area......	68·6 sq. cms.
Time of run.................................	1 hr. 3 min.
Average current through bath................	0·344 amp.
Theoretical weight of nickel deposited.........	0·378 grams.
Weight of cathode before plating.............	60·5166 "
Weight of cathode after plating..............	60·8678 "
Nickel deposited...........................	0·3512 "

$$\text{CURRENT EFFICIENCY} \quad \frac{0·351}{0·378} = 93·0\%$$

SOLUTION VI C.

Cobalt sulphate $CoSO_4$.................. 38·6 grams.
Ammonium sulphate $(NH_4)_2SO_4$......... 32·9 "
Boric acid........................... 38·8 "
Water...............................1000 c.c.
　　　Total bath...................... 2·5 litres.

No addition of reagents was made subsequent to the making of the original solution, either to change its reaction or for other purpose.

Data of Plating Experiments, Solution VI C.

			Cast Anodes.			
Date of experiment.	Current density in amperes per sq. decimetre.	Electromotive force in volts.	Electrode distance in cms.	Cathode area in sq. cms.	Duration of plating.	Character of plate and remarks.
1914 July 20	0·50	0·94	10	Brass 27·3	2　0	A very smooth, bright satisfactory plate when taken from solution, which required very little buffing to finish.
July 23	0·70	1·12	10	Brass 27·3	2　0	Plate was a dull metallic colour when taken from bath but buffed up to satisfactory surface. Plate showed few pits near bottom.
Aug. 4	1·0	1·95	10	Brass 39·3	1　30	Plate grey in colour as removed from bath but readily buffed to satisfactory finish.
Aug. 5	1·5	. 2·15	10	Brass 27·3	2　0	Plate dull in colour and burned in places.

Current Efficiency Test, Solution VI C.

Run I.

Current density approx...................... 0·50 amp. per sq. dec.
Cathode—brass polished both sides—area....... 72·0 sq. cms.
Time of run............................... 1 hr. 0 min.
Average current through bath.............. 0·357 amp.
Theoretical weight of cobalt deposited........ 0·393 grams.
Weight of cathode before plating............. 59·8936 "
Weight of cathode after plating.............. 60·2659 "
　　　　　　　　　　　　　　　　　　　　　　　—————
Cobalt deposited........................... 0·3723 "

$$\text{CURRENT EFFICIENCY} \quad \frac{0·372}{0·393} = 94·6 \%$$

Run II.

Current density approx...................... 0·50 amp. per sq. dec.
Cathode—brass polished both sides—area...... 72·0 sq. cms.
Time of run............................... 1 hr. 0 min.
Average current through bath.............. 0·361　amp.
Theoretical weight of cobalt deposited........ 0·397 grams.
Weight of cathode before plating............. 62·7582 "
Weight of cathode after plating.............. 63·1299 "
　　　　　　　　　　　　　　　　　　　　　　　—————
Cobalt deposited........................... 0·3717 "

$$\text{CURRENT EFFICIENCY} \quad \frac{0·372}{0·397} = 93·7 \%$$

Run III.

Current density approx. . . . . 1·0 amp. per sq. dec.
Cathode—brass polished both sides—area. . . . . . 39·2 sq. cms.
Time of run. . . . . 1 hr. 0 min.
Average current through bath. . . . . 0·394 amp.
Theoretical weight of cobalt deposited. . . . . 0·433 grams.
Weight of cathode before plating. . . . . 29·6882 "
Weight of cathode after plating. . . . . 30·0979 "

Cobalt deposited. . . . . 0·4097 "

$$\text{CURRENT EFFICIENCY} \quad \frac{0·410}{0·433} = 94·6\%$$

Run IV.

Current density approx. . . . . 1·0 amp. per sq. dec.
Cathode—brass polished both sides—area. . . . . . 40·0 sq. cms.
Time of run. . . . . 1 hr. 0 min.
Average current through bath. . . . . 0·402 amp.
Theoretical weight of cobalt deposited. . . . . 0·442 grams.
Weight of cathode before plating. . . . . 37·7798 "
Weight of cathode after plating. . . . . 38·1982 "

Cobalt deposited. . . . . 0·4184

$$\text{CURRENT EFFICIENCY} \quad \frac{0·418}{0·442} = 94·6 \%$$

SOLUTION VI D

This solution is the same as I B, except that boric acid is added to it. In fact, the identical solution I B was used with boric acid addition. The final composition of the bath was as follows:—

Cobalt-ammonium sulphate $CoSO_4(NH_4)_2SO_4·6H_2O$ 200 grams.
Boric acid. . . . . .37·2 "
Water. . . . . .1000 c.c.

In making up this solution, after the boric acid was thoroughly dissolved by continued agitation, it was allowed to stand for several days to see if any salt would crystallize. The solution was perfectly clear after this period, and appeared to be ready for use as a plating bath.

Data of Plating Experiments, Solution VI D.

Date of experiment.	Current density in amperes per sq. decimetre.	Electromotive force in volts.	Electrode distance in cms.	Cathode area in sq. cms.	Duration of plating.		Character of plate and remarks.
1914 Sept. 25	0·51	1·0	10	Brass 35·5	1	0	Good, even uniform deposit. Grey when taken from bath, but showing good lustre when buffed.
Sept. 25	0·75	1·35	10	Brass 34·1	1	30	Smooth and uniform deposit, showing good lustre when polished.
Sept. 26	1·0	2·85	10	Brass 34·3	1	0	Good, smooth uniform grey deposit, taking good lustre when polished.
Oct. 1	1·0	1·25	10	34·3	1	0	Same
Oct. 1	1·25	1·4	10	Brass 32·6	1	0	Same
Oct. 1	1·25	1·38	10	Brass 34·6		0	Same
Sept. 28	1·5	1·35	10	Brass 35·3	–	0	Uniform dull white deposit, showing tendency to split at the bottom and edges.
Sept. 28	1·5	1·92	10	Brass 37·5	1	30	Uniform smooth grey plate, showing tendency to peel at the edges.
Sept. 29	1·5	1·87	10	Brass 33·8	1	0	Uniform light grey deposit, split at edges.
Oct. 2	1·5	1·74	10	33·8	1	0	Uniform rough, grey porous deposit, impossible to buff.
Oct. 2	1·75	2·11	10	Brass 17·8	0	30	Uniform grey deposit, badly gas pitted.

CONCLUSIONS.

(1)   Solution VI A does not yield a satisfactory plate at any current density, plating with cast anodes, up to 1·0 amperes per square decimetre. The plates are dark and spotted and hydrogen is evolved. At current densities in the neighbourhood of one ampere per square decimetre or higher, the plates are badly burned and pitted.   With rolled anodes there is a slight improvement at moderate current densities, that is, in the neighbourhood of 0·50 amperes per square decimetre.

(2)   Solution VI B, which is the nickel analogue of VI A gave very bright, clean satisfactory plates at current densities up to 0·50 amperes per square decimeter.   At current densities greater than this the plates are burned.

(3)   This nickel solution VI B is not nearly so fast as some of the cobalt solutions elsewhere described in this paper.

(4)   Solution VI C, which is a more concentrated solution of the type of VI A, operated satisfactorily at a higher current density than VI A, that is up to 1·0 ampere per square decimetre.   This conforms to the general conclusion throughout this work that the more concentrated cobalt solutions, which are likewise those from which plates may be obtained with the greatest speed, are the most satisfactory for practical plating purposes.

(5)   Solution VI D is not to be compared in speed with I B, from which it differs only by the addition of the boric acid.   Solution VI D yields a satisfactory white cobalt plate at all current densities up to 1·25 amperes per square decimetre.

(6)   The current efficiencies with all solutions of Series 6 are high and satisfactory, the average being better than 90 per cent.

(7)   These solutions operate alike, plating on iron, steel and brass, for which metals only the above conclusions apply.

(8)   The cobalt plates from Series 6, wherever they are pronounced satisfactory, are firm, adherent, hard and uniform, and may readily be buffed to a satisfactorily finished surface. They take a very high polish with a beautiful lustre, which, although brilliantly white, possesses a slightly bluish cast.

SERIES 7: COBALT-AMMONIUM SULPHATE, COBALT CARBONATE AND BORIC ACID SOLUTIONS.

This solution was prepared by boiling cobalt-ammonium sulphate crystals and cobalt carbonate in water until the evolution of CO_2 gas ceased, and until the solution was no longer acid to blue litmus. After settling, any undissolved cobalt carbonate was removed by filtration or decantation, and the boric acid added. After further boiling for a short time, the solution was allowed to cool and was ready for plating purposes. If the solution is too acid it may be neutralized with ammonia.

Solution VII A can be used with rolled anodes, or rolled anodes with a small proportion of cast ones. The proportion of cast to rolled anodes used in this bath can be established by frequent testing of the acidity of the bath. For this purpose the following should be noted. Blue litmus paper should always be reddened indicating the presence of boric acid, congo paper should not be turned blue, for if it does, it indicates the presence of free sulphuric acid in the bath. Red litmus paper should remain red for if it turns blue the bath has become alkaline and further addition of boric acid is required.

After plates from solution VII A were found to be unsatisfactory, yielding dark deposits at current densities in the neighbourhood of 1 ampere per square decimetre, sodium sulphite was added to the bath to see if the deposit would be brighter. This latter solution is solution VII B.

SOLUTION VII A.

Cobalt-ammonium sulphate $CoSO_4 \cdot (NH_4)_2SO_4 \cdot 6H_2O$. . 63·5 grams.
Cobalt carbonate $CoCO_3$. . . . . 5·3 "
Boric acid H_3BO_3. . . . . .31·7 "
Water. . . . . .1000 c.c.
 Total bath. . . . . 3·5 litres.

Data of Plating Experiments, Solution VII A.

				Cast Anodes.		
Date of experiment.	Current density in amperes per sq. decimetre.	Electro-motive force in volts.	Electrode distance in cms.	Cathode area in sq. cms.	Duration of plating	Character of plate and remarks.
July 3	0·35		20	Brass 27·3	2 0	Bright even plate without flaw, which buffed satisfactorily.
July 3	0·50	1·50	20	Brass 27·3	2 0	Very bright smooth plate without flaws, buffed brightly.
July 9	0·50	1·55	20	Brass 39·3	2 30	Surface granular. Unsatisfactory.
July 10	0·50	1·53	20	Brass 39·3	2 30	Plate rough and unsatisfactory, and did not polish readily.
July 4	0·70	2·7	20	Brass 27·3	2 0	Smooth plate, free from flaws, white but not lustrous. Buffed satisfactorily.
July 6	1·0	2·5	10	Brass 27·3	2 0	Plate burned at sides and bottom. unsatisfactory. Current density too great.
July 7	1·0	2·7	15	Brass 27·3	2 0	Plates burned at edges, unsatisfactory, Current density too great.
July 7	1·0	3·30	15	Brass 39·3	2 30	Heavy deposit, but dull and granular. Burned.
July 9	1·0	3·25	15	Brass 39·3	2 30	Plate burned and unsatisfactory.
July 11	1·0	4·15	15	Brass 27·3	2 0	Plate scaley and unsatisfactory.

Solution rapidly becoming more alkaline with use.

SOLUTION VII B.

Cobalt-ammonium sulphate $CoSO_4 \cdot (NH_4)_2SO_4 \cdot 6H_2O$ 63·5 grams.
Cobalt carbonate $CoCO_3$.......................... 5·3   "
Boric acid....................................... 31·7   "
Sodium sulphite Na_2SO_3.......................... 1·4   "
Water... 1000 c.c.
    Total bath................................. 3·5 litres.

Data of Plating Experiments, Solution VII B.

			Cast Anode.			
Date of experiment.	Current density in amperes per sq. decimetre.	Electromotive force in volts.	Electrode distance in cms.	Cathode area in sq. cms.	Duration of plating.	Character of plate and remarks.
July 11	1·0	2·05	15	Brass 27·3	2 0	White even plate, somewhat burned at edges.
			Solution rapidly becoming more alkaline.			
			Rolled Anode.			
July 17	0·50	1·21	10	Brass 27·3	2 0	An even satisfactory plate.
July 17	0·50	1·15	10	Brass 27·3	2 0	Satisfactory deposit, bright and required but little buffing.
July 17	0·80	1·95	10	Brass 27·3	2 0	Satisfactory bright plate.
July 18	0·90	2·20	10	Brass 27·3	2 0	Black burned plate.

Conclusions.

(1)   Solution VII A gives bright, satisfactory plates, at low current densities in the neighbourhood of 0·35 amperes per square decimetre, as is true of the corresponding nickel solution.

(2)   Solution VII A does not lend itself for rapid plating; at current densities over 0·80 amperes per square decimetre, the plates are burned.

(3)   The addition of sodium sulphite to solution VII A, as in solution VII B, does not materially increase the current density at which satisfactory bright plates may be obtained.   These solutions are not to be compared for rapid plating with some of the others described and styled satisfactory, as I B and XIII B.

(4)   The solutions of Series 7 become alkaline so rapidly with cast anodes, that rolled anodes should preferably be used with them.   In general, cobalt solutions improved by addition of sodium sulphite are not sufficiently constant with prolonged use to have the self supporting characteristic required by most plating establishments.

(5)   The solutions of Series 7 operated alike, plating on iron, steel and brass, for which metals only the above conclusions apply.

(6)   The cobalt plates from Series 7, wherever they are pronounced satisfactory, are firm, adherent, hard and uniform, and may readily be buffed to a satisfactory finished surface.   They take a very high polish, with a beautiful lustre, which, although brilliantly white, possesses a slightly bluish cast.

Series 8: Cobalt Sulphate, Potassium Citrate and Ammonium Chloride Solutions.

This bath is analogous to one recommended by Langbein,[1] C. H. Proctor,[2] and W. Pfanhauser,[3] and which is reported to be particularly

---

[1] Dr. Geo. Langbein, "Electrodeposition of Metals" 6th Edition, Revised.
[2] Metal Industry, 1911, p. 353.
[3] W. Pfanhauser, Elektroplatticrung, 1900.

satisfactory for plating on copper and zinc. Langbein specifies for the nickel solution on copper and copper alloys, a current density of $0\cdot45$ amperes per square decimetre, and on zinc a current density of $0\cdot8$ to 1 ampere per square decimetre.

SOLUTION VIII A

Cobalt sulphate $CoSO_4$..................... 26·3 grams.
Potassium citrate........................ 17·6 "
Ammonium chloride NH_4Cl.............. 17·6 "
Water................................. 1000 c.c.
Total bath........................... 2·5 litres

Data of Plating Experiments, Solution VIII A.

				Cast Anodes.			
Date of experiment.	Current density in amperes per sq. decimetre.	Electromotive force in volts.	Electrode distance in cms.	Cathode area in sq. cms.	Duration of plating.		Character of plates and remarks.
1914 Oct. 1	0·30	0·67	10	Brass 34·8	2	0	Good, uniform bluish white deposit, buffed readily to mirror surface.
Oct. 1	0·50	0·97	10	Brass 33·1	2	0	Same
June 30	0·50	1·95	20	Brass 27·3	3	0	Beautiful bright hard, smooth plate after polishing.
July 15	0·50	1·15	13	Brass 37·1	1	30	Dark plate on removal from solution, but buffed up satisfactorily.
Oct. 2	0·95	1·50	10	Brass 40·8	1	30	Smooth, uniform, grey deposit, readily buffed to mirror surface.
Oct. 2	1·0	1·36	10	Brass 34·1	1	0	Smooth, uniform, grey deposit, buffed to satisfactory finish.
Oct. 15	1·5	2·4	10	Brass 35·3	1	0	Uniform, grey deposit showing burning.
Oct. 15	1·75	2·0	10	Brass 30·6	0	45	Same

Solution VIII B differs from solution VIII A only in that it is more concentrated in $CoSO_4$.

SOLUTION VIII B.

Cobalt sulphate $CoSO_4$..................... 40·5 grams.
Potassium citrate........................ 17·6 "
Ammonium chloride $(NH_4) Cl$.......... 17·6 "
Water................................. 1000 c.c.
Total bath........................ 1·0 litres.

Data of Plating Experiments, Solution VIII B.

Date of experiment.	Current density in amperes per sq. decimetre.	Electro-motive force in volts.	Electrode distance in cms.	Cathode area in sq. cms.	Duration of plating.	Character of plate and remarks.
Sept. 22	0·26	0·95	10	Cast zinc 26·6	2 0	Crystalline in most parts, but with small patches of smooth plate.
Oct. 2	0·50	1·07	10	Brass 39·6	2 0	Uniform, smooth, grey deposit, readily buffed to mirror surface.
Sept. 22	1·0	1·97	10	Brass 34·7	1 0	Good, smooth, uniform plate, somewhat dark, but taking a good polish when buffed.
Oct. 1	1·0	1·4	10	Brass 34·7	1 0	Good, smooth, uniform plate, somewhat dark, but taking a good polish when buffed.
Oct. 1	1·25	1·75	10	Brass 33·4	1 0	Uniform smooth, white plate, buffing to mirror surface.
Sept. 23	1·5	2·67	10	Brass 35·0	30	Good, smooth, uniform deposit, dark grey, but taking a good polish. Slightly burned at edges.
Oct. 1	1·5	2·15	10	Brass 34·8	1 0	Uniform, smooth, white deposit, slightly pitted at top and burned at edges. Not readily buffed.
Oct. 2	1·75	2·35	10	Brass 34·8	1 0	Smooth, uniform grey deposit, slightly burned on edges and difficult to buff.
				Zinc Cathode.		
Oct. 5	0·20	0·80	10	Cast zinc 26·6	3 0	Smooth, uniform metallic dark grey plate. Buffed readily to mirror surface.
Oct. 10	0·50	1·03	10	Same	2 0	Same. Mechanical agitation of solution near cathode.
Oct. 6	0·75	0·95	10	Cast zinc 25·6	3 0	Very rough, scaley, unsatisfactory deposit. Mechanical agitation.
Sept. 21	1·0	1·4	10	Cast zinc 26·6	1 0	Poor plate, very decidedly crystalline deposit. No agitation.

CONCLUSIONS.

(1) Cobalt plates from these cobalt sulphate solutions containing potassium citrate and ammonium chloride, on brass and iron, are firm, adherent, hard and uniform and may readily be buffed to a satisfactorily finished surface. They take a very high polish, with a beautiful lustre, which, although brilliantly white, possesses a slightly bluish cast.

(2) Neither solution VIII A nor solution VIII B lends itself to fast cobalt plating like solutions I B and XIII B. Solution VIII A yields satisfactory deposits at all current densities up to 1·0 ampere per square decimetre, while solution VIII B, which is more concentrated in cobalt sulphate, yields satisfactory plates at all concentrations up to 1·25 amperes per square decimetre. These figures are for plating on brass and iron.

(3) Solution VIII B may be used for plating on zinc at low current densities up to 0·50 amperes per square decimetre, particularly if the solution near the cathode is agitated.

SERIES 9: COBALT-PHOSPHATE, SODIUM-PYROPHOSPHATE SOLUTIONS.

SOLUTION IX.

Cobalt phosphate $CoHPO_4$	7·58 grams.
Sodium-pyrophosphate	66·1 "
Water	1000 c.c.
Total bath	3·5 litres.

Langbein[1] recommends a solution containing nickel phosphate 15·8 grams, sodium-pyrophosphate 66·1 grams and water 1000 c.c. for dark nickelling upon iron, brass and copper. This is supposed to be particularly serviceable where darker tones of nickel are required for decorative purposes.

It was found in trying to prepare the analogous cobalt phosphate solution that the solubility of the cobalt phosphate was lower than that of nickel phosphate. Solution IX is saturated in cobalt phosphate.

The cobalt phosphate prepared for this bath was made by mixing two solutions one containing 30·0 grams $CoSO_4 \cdot 7H_2O$ in 3·4 litres of warm water, and the other containing 24·9 grams sodium phosphate in 3·4 litres of warm water. These two solutions were mixed with constant stirring, and the precipitated cobalt phosphate filtered off. These quantities yielded 15·8 grams cobalt phosphate.

The final bath was prepared by dissolving the sodium pyrophosphate in warm water, and adding the cobalt phosphate, which dissolved up to the quantity indicated in solution IX as determined by analysis but not up to the quantity recommended by Langbein for nickel.

Data of Plating Experiments, Solution IX.

						Cast Anodes.
Date of experiment.	Current density in amperes per sq. decimetre.	Electromotive force in volts.	Electrode distance in cms.	Cathode area in sq. cms.	Duration of plating.	Character of plate and remarks.
1914 July 3	0·50	3·30	10	Brass 27·3	1 30	Plate coated with precipitate after removal from solution, which was easily washed off. Satisfactory and readily buffed.
Sept. 29	0·50	3·03	10	Brass 33·4	2 0	Dark streaked deposit. Gelatinous cobalt compound precipitated on surface of cathode.
Sept. 30	0·75	3·68	10	Brass 34·1	2 0	Dark, lustrous, streaked deposit. Gelatinous precipitate as in last.
July 4	0·80	3·65	10	Brass 27·3	2 0	A bright satisfactory plate as taken from solution, except for few black spots, which buffed off easily.
July 6	1·0	4·80	10	Brass 27·3	1 30	Smooth, even plate, which was covered with bluish precipitate when removed from bath. Satisfactory upon buffing.
July 7	1·0	5·0	20	Brass 27·3	1 45	Black, streaked and unsatisfactory.
Sept. 30	1·0	4·35	10	Brass 40·0	1 30	Dark, streaked and unsatisfactory. Gelatinous precipitate on cathode.

CONCLUSIONS.

(1) Solution IX is not more satisfactory than the corresponding nickel phosphate solution for the purpose of dark cobalting for decorative purposes. The voltage required for moderate current densities is extremely high as compared with that required for other cobalt solutions described as satisfactory.

(2) Cobalt phosphate, in the presence of sodium pyrophosphate, as indicated in solution IX, is less soluble than the corresponding nickel salt. This is the only case among those studied in which the cobalt bath could not be made more concentrated in metallic cobalt than the corresponding nickel bath, with consequent greater electrical conductivity and correspondingly higher permissible current density for plating.

---

[1] Dr. Geo. Langbein, Electro-deposition of Metals, 6th Edition revised, p. 258.

SERIES 10: COBALT-AMMONIUM-SULPHATE MAGNESIUM SOLUTIONS.

SOLUTION X.

Cobalt sulphate	26·6 grams.
Ammonium sulphate	22·6 "
Magnesium sulphate	33·8 "
Water	1000 c.c.
Total bath	2·5 litres.

Data of Plating Experiments, Solution X.

Date of experiment.	Current density in amperes per sq. decimetre.	Electromotive force in volts.	Electrode distance in cms.	Cathode area in sq. cms.	Duration of plating.		Character of plate and remarks.
1914 July 29	0·30	0·70	10	Brass 39·3	1	30	Plate of white metallic colour as taken from solution, and easily buffed to mirror surface.
Sept. 29	0·30	0·75	10	Brass 36·6	2	0	Good, white uniform deposit requiring but little buffing to bring to mirror surface.
Sept. 30	0·40	0·87	10	Brass 32·2	2	0	Same
Sept. 30	0·50	1·03	10	Brass 35·2	1	30	Same.
July 6	0·50	1·54	15	Brass 48·1	2	0	A bright plate, evenly coated except for few streaks in metal portion. Readily buffed to satisfactory mirror surface.
July 16	0·50	1·1	14	Brass 37·1	1	30	Bright, uniform plate obtained over entire surface. Readily buffed to satisfactory mirror surface.
July 29	0·50	1·15	10	Brass 46·6	2	0	Plate of white metallic colour as taken from solution, and easily buffed to mirror surface.
Sept. 30	0·60	1·17	10	Brass 34·3	1	30	Bright, uniform plate obtained over entire surface. Readily buffed to satisfactory mirror surface.
July 7	0·70	1·70	15	Brass 48·1	2	0	Bright plate, evenly coated, except for few streaks in middle portion. Readily buffed to satisfactory mirror surface.
Sept. 30	0·70	1·34	10	Brass 34·8	1	30	Good, uniform white deposit, requiring but little buffing to bring to mirror surface.
July 7	0·70	1·30	15	Brass 48·1	1	30	Satisfactory plate, white, smooth and even. Readily buffed to satisfactory mirror surface.
July 15	0·90	2·25	15	Brass 37·1	1	0	Plate bright and smooth at centre, but somewhat burned at the edges.
July 7	1·0	2·25	15	Brass 48·1	1	30	Bright plate, rough and uneven on lower portion, no burning apparent.
July 7	1·0	2·05	15	Brass 39·3	15	0	A very thick plate, which was rough, dark in colour, full of small holes, but adherent.
July 9	1·0	2·10	15	Brass 39·3	3	0	Pitted and somewhat burned.
Rolled Anodes.							
July 16	0·5	1·0	14	Brass 37·1	2	0	Bright, smooth, uniform plate over entire surface.
July 16	0·8	1·05	15	Brass 37·1	1	0	Bright, smooth plate over entire surface.
July 17	1·0	2·25	17	Brass 37·1	1	0	Plate was fairly smooth and bright, but not very uniform in appearance, some parts darker than others.
July 17	1·0	2·15	17	Brass 37·1	1	0	Plate very smooth, bright and uniform.

CONCLUSIONS.

(1) Cobalt plates from solution X on brass and iron are firm, adherent, hard and uniform, and may readily be buffed to satisfactory mirror surface. They take a very high polish with a beautiful lustre, which although brilliantly white, possesses a slightly bluish cast.

(2) The soft yellowish tinge which is observed when plating with the nickel analogue of solution X, was not found with the cobalt solution. On the contrary the plates are beautifully white and hard.

(3) The specific electrical conductivity of solution X is very much higher than that of the corresponding nickel solution.

(4) All of the cobalt plates deposited at current densities between 0·25 and 0·75 are as smooth, adhesive, and generally satisfactory as the best nickel plates.

(5) Solution X does not lend itself to extremely fast plating like solutions I B and XIII B but satisfactory plates may be obtained with it at current densities up to 0·75 amperes per square decimetre. Solution X may be used at very much higher current densities than the corresponding nickel solution, for which a current density of 0·20 amperes per square decimetre is recommended.

SERIES 11: COBALT SULPHATE, AMMONIUM-TARTRATE, TANNIC ACID SOLUTIONS.

SOLUTION XI.

$$
\begin{aligned}
&\text{Cobalt sulphate } CoSO_4 \dots\dots\dots\dots\dots\dots\dots \quad 25\cdot0 \text{ grams.}\\
&\text{Ammonium-tartrate } (NH_4)_2C_4H_4O_6 \dots\dots\dots \quad 41\cdot7 \quad \text{“}\\
&\text{Tannic acid } H\ C_4H_9O_9 \dots\dots\dots\dots\dots\dots \quad 0\cdot28 \quad \text{“}\\
&\text{Water} \dots\dots\dots\dots\dots\dots\dots\dots\dots\dots\dots\dots \quad 1000 \text{ c.c.}\\
&\qquad\text{Total bath} \dots\dots\dots\dots\dots\dots\dots \quad 3\cdot5 \text{ litres.}
\end{aligned}
$$

Neutral ammonium-tartrate is obtained by saturating a solution of tartaric acid with ammonia. The cobalt salt should also be neutral. The solution is prepared by dissolving the ingredients shown above in water, boiling for about fifteen minutes, adding water to make desired quantity, and filtering.

Data of Plating Experiments, Solution XI.

Date of experiment.	Current density in amperes per sq. decimetre.	Electromotive force in volts.	Electrode distance in cms.	Cathode area in sq. cms.	Duration of plating.		Character of plate and remarks.
1914 July 7	0·30	1·5	25	Brass 41·2	1	30	Cathode covered with a dark deposit which was absolutely unsatisfactory.

This solution continually precipitated a cobalt compound both when in use and upon standing idle. It gave the same unsatisfactory deposit as in the run shown above, at the several current densities tried.

CONCLUSION.

Solution XI is not satisfactory for cobalt plating under the usual conditions of plating practice.

SERIES 12: COBALT SULPHATE, POTASSIUM HYDRATE, AND TARTARIC ACID SOLUTIONS.

SOLUTION XII.

Cobalt sulphate CoSO₄.................	52·7 grams.
Tartaric acid H₂C₄H₄O₆.................	27·8 "
Caustic potash KOH..................	6·8 "
Water...............................	1000 c.c.
Total bath......................	3·5 litres.

The cobalt sulphate, tartaric acid and caustic potash were dissolved in water and then mixed, adding sufficient water to make the above bath.

Data of Plating Experiments, Solution XII.

				Cast Anodes.		
Date of experiment.	Current density in amperes per sq. decimetre.	Electromotive force in volts.	Electrode distance in cms.	Cathode area in sq. cms.	Duration of plating.	Character of plate and remarks.
1914 July 6	0·30	1·17	20	Brass 37·1	2 0	Plate white, uniform and velvety. Readily buffed to mirror surface.
Oct. 1	0·30	1·25	10	Brass 39·5	2 0	Same
Sept. 29	0·39	1·25	10	Brass 33·6	2 0	Smooth uniform dark deposit, which buffed to satisfactory mirror surface.
Sept. 30	0·50	1·45	10	Brass 34·6	2 0	Uniform smooth, grey deposit, which buffed to satisfactory finish.
Oct. 1	0·50	1·40	10	Brass 34·8	2 0	Same
Oct. 2	0·50	1·53	10	Brass 41·2	2 0	Same
Sept. 30	0·75	2·03	10	Brass 39·1	1 30	Uniform, smooth grey deposit, which buffed satisfactorily.
July 13	1·0	3·6	15	Brass 37·0	1 30	A very even smooth satisfactory plate.
Oct. 2	1·0	2·18	10	Brass 32·6	1 0	Same
Oct. 2	1·25	2·55	10	Brass 32·9	1 30	Same
Oct. 3	1·50	2·86	10	Brass 30·9	0 45	Same
Oct. 15	1·50	3·5	10	Brass 34·6	1 0	Same
Oct. 15	1·96	4·25	10	Brass 30·7	0 30	A good, smooth, white plate, nearly polished as removed from solution.
Oct. 12	3·87	6·00	10	Brass 34·0	0 15	Same
Oct. 12	4·0	3·95	10	Brass 14·2	0 30	Same
Oct. 15	5·0		10	Brass 13·0	0 20	Same
Oct. 16	6·0		10	Same	0 15	Same
Oct. 16	7·0	6·0	10	Brass 9·9	0 10	Same
Oct. 17	8·0	4·9	10	Brass 10·1	0 5	Same
Oct. 17	10·0	6·0	10	Brass 10·6	0 7	Same
Oct. 17	12·0	5·5	10	Brass 11·5	0 5	Plate burned at edges.
Oct. 17	15·6	5·8	10	Brass 9·3	0 5	Badly burned at edges—splitting.
				Rolled Anodes.		
July 16	1·0	3·0	14	Brass 37·0	1 30	Plate smooth and uniform, but dark. The solution from which this plate was taken had been operated for 48 hours.

CONCLUSIONS.

(1) Cobalt plates from solution XII, which is simple cobalt sulphate in the presence of potassium tartrate with an excess of tartaric acid, on brass and iron, are firm, adherent hard and uniform and may readily be buffed to a satisfactorily finished surface. They take a very high polish with a beautiful lustre, which, although brilliantly white, possesses a slightly bluish cast.

(2) All of the cobalt plates from this solution within the current density ranges described as satisfactory, are as smooth, adhesive and generally satisfactory as the best nickel plates.

(3) Solution XII is an extremely fast plating solution when compared with the fastest nickel solutions. It yields satisfactory plates at all current densities up to $11 \cdot 0$ amperes per square decimetre.

(4) Solution XII may be used for plating on brass, iron and steel, for which cathodes alone these conclusions apply.

(5) There is no nickel bath, of which we are aware, operating in the manner of the usual plating practice at anything like as high a current density as the cobalt solution XII.

SERIES 13: COBALT SULPHATE, COBALT CHLORIDE AND BORIC ACID SOLUTIONS.

Solution XIII A is analogous to one suggested to us by Mr. W. S. Barrows, foreman of the plating department, Russel Motor Car Company, Toronto, Ontario, as being satisfactory and rapid for nickel plating. The cobalt analogue was made up as follows:—

SOLUTION XIII A.

Cobalt sulphate $CoSO_4$	$181 \cdot 2$ grams.
Sodium chloride NaCl	$11 \cdot 35$ "
Boric acid	$37 \cdot 8$ "
Water	1000 c.c.
Total bath	4 litres.

This solution is not nearly saturated in cobalt sulphate.

Data of Plating Experiments, Solution XIII A.

				Cast Anodes.			
Date of experiment.	Current density in amperes per sq. decimetre.	Electromotive force in volts.	Electrode distance in cms.	Cathode area in sq. cms.	Duration of plating.		Character of plate and remarks.
1914 July 10	0·70	1·56	23	Brass 37·1	1	30	Plate rather dark as removed from bath. Required but little buffing to finish satisfactorily, greatly resembled Ni in colour.
July 10	0·70	1·56	23	Brass 37·1	1	30	Same
July 10	0·97	2·0	15	Brass 37·1	2	0	Deposit fairly smooth and uniform but dark.
July 11	1·0	1·8	16	Brass 37·1	2	0	Same
July 11	1·0	1·87	15	Brass 37·1	1	0	Good smooth, uniform bright deposit, but somewhat streaked on upper half of plate.
July 14	1·0	1·45	10	Brass 37·1	1	30	Plate very smooth and uniform on lower half, but badly streaked in upper portion.
July 14	1·0	1·85	15	Brass 37·1	1	0	Same
July 14	1·0	1·7	13	Brass 37·1	1	0	Solution was stirred continuously by bubbling air through near cathode. Plate brighter and more metallic looking than last runs, but badly split and peeled.
July 15	1·0	1·9	12	Brass 37·1	0	45	Plate very poor being dark streaked and peeling.

SOLUTION XIII B.

Cobalt sulphate CoSO$_4$.................. 312·5 grams.
Sodium chloride NaCl.................. 19·6 "
Boric acid.......................... nearly to saturation.
Water.............................. 1000 c.c.
Total bath approximately........... 1½ litres.

This solution is substantially saturated in cobalt sulphate in the presence of the other components. Its specific gravity at 15°C is approximately 1·24.

Data of Plating Experiments Solution, XIII B.

Date of experiment.	Current density in amperes per sq. decimetre.	Electromotive force in volts.	Electrode distance in cms.	Cathode area in sq. cms.	Duration of plating.	Character of plate and remarks.
Oct. 2	0·50	1·02	10	Brass 33·6	3 0	Uniform rough dark plate. Impossible to polish without grinding.
Oct. 2	0·75	1·23	10	Brass 32·6	1 30	Smooth, uniform grey plate, which buffed to satisfactory finish with difficulty.
Oct. 3	1·0	1·53	10	Brass 34·0	1 0	Uniform, dark grey deposit, difficult to buff.
Oct. 5	1·25	1·75	10	Same	Same	Same
Oct. 6	1·50	1·25	10	Brass 32·0	Same	Uniform smooth grey plate, which buffed somewhat more readily than preceding plates with this solution.
Oct. 6	5·46	5·5	10	Brass 34·8	0 30	Good, smooth, white deposit, which buffed readily to mirror surface.
Oct. 6	6·0		10	Brass 32·7	0 10	Good, smooth, white deposit, buffed readily to mirror surface.
Oct. 10 After severe ageing test described below.	6·15	6·0	10	Brass 32·7	0 15	Same
Oct. 8	8·0	4·83	10	Brass 17·6	0 12	Same
Oct. 9	8·0	6·0	10	Iron 20·0	0 10	Good smooth, white deposit, which buffed readily to mirror surface. This plate was given a severe bending test, being doubled on itself backwards and forwards to an angle of 180 degrees. The metal furrowed and split on surface and end, but plate clung absolutely.
Same	Same	Same	Same	Same	Same	Same
Oct. 8	8·8	6·5	10	Iron 22·5	0 12	Good, smooth, white deposit, which buffed readily to mirror surface.
Oct. 10 after severe ageing test	8·88	5·90	10	Brass 20·0	0 15	Good, smooth, white deposit, which buffed readily to mirror surface.
Oct. 8	9·77	6·5	10	Iron 22·5	0 12	Same
Oct. 15	10·7	6·3	10	Iron 1·81	0 20	Same
Oct. 8	10·0		10	Brass 17·6	0 5	Same
Oct. 8	14·6	6·5	10	Brass 14·7	0 5	Same Best deposit obtained with this solution to date, although all plates at current densities from 6 amperes up were good.
Oct. 13	16·5	5·7	10	Brass 10	0 5	Good, smooth white deposit, readily buffed to mirror surface. Excellent plate.

Continued

Oct. 6	17·5	6·85	10	Brass 10·6 *Heavy Plate*	0	5	Same
Oct. 6-7	5·35	6·0	10	Brass 32·8	15	15	Firm, adherent, massive plate, showing no tendency to split or curl. Smooth in centre, with modules at edges. Weight approximately 37 grams, thickness approximately 1 mm.

The area of the effective cathode increased from 32·8 sq. cms. at the start, to approximately 40·0 sq. cms. at the end of the run. This latter figure is not sufficiently accurate to admit of exact computation, but the figures show in a general way that the current efficiency was very high.

Oct. 7	16·5	5·5	10	Brass 10·0	17	30	Firm, adherent, massive plate showing no tendency to split or curl. Weight about 30 grams.
Oct. 13	5·26	6·0	10	Brass circular 18·9	67	0	Firm, adherent, massive plate showing no tendency to split or curl; about 5 mms. thick.

Plate on brass cathodes with grooves depth 1·62 mms. to 7·0 mms. to study "throwing" property of this solution.

Oct. 7	9·0	5·5	10	Brass 18·3	0	5	Good, smooth, uniform white deposit. All grooves satisfactorily covered, and satisfactorily buffed to mirror finish.
Oct. 7	3·77	5·5	10	Brass Block 60·1 total surface	0	10	Same

Grooves faced toward anode, but entire block, back as well as grooves satisfactorily covered to admit of severe and satisfactory buffing after only 10 minutes plating.

Oct. 7	3·83	5·5	10	Brass Block 60·1 total surface	0	20	Same

Same remark as previous plate.

Further Data of Plating Experiments with Cobalt Solution XIII B.

Experiments at Russel Motor Car Company Plating Plant.

Solution:—

Water—4¾ gals.   Boric acid—3 lbs.   Acidity—strongly acid.
NaCl—15½ ozs.   Cobalt-Sulphate crystals   Sp. Gr.—28·5 Be.
        26 lbs., 10½ ozs.   Temperature 64.

Date of experiment.	Current density in amperes per sq. decimetre.	Electromotive force in volts.	Electrode distance in ins.	Cathode area in sq. ins.	Duration of plating.		Character of plate and remarks
					hr.	min.	
Nov. 2	6·5	4·25	5·0	Brass 12	0	5	Smooth, hard, firm plate, very adherent; stood bending test and twisting. Buffed on 12″ cotton buff wheel at 3600 R.P.M. with pressure approximately same as for heavy nickel plate. No evidence of cutting through; splendid results.
Nov. 2	4·4	3·5	5·0	Steel 28	1	0	Mottled burnt deposit, impossible to buff or polish to a satisfactory surface. Solution full of floating particles of boric acid.
Nov. 2	3·2	3·5	5	Brass 42·5	1	0	Mottled cloudy plate at edges, centre quite brilliant, but difficult to colour. Poor plate.
		(*Filtered* *Solution*)					

Continued.

Date of experiment.	Current density in amperes per sq. decimetre.	Electromotive force in volts.	Electrode distance in inches.	Cathode area in sq. inches.	Duration of plating.		Character of plate and remarks.
Nov. 5	3·5	5	5	Brass Hub Cap 70	0	30	Very heavy, dirty grey plate, flaked on sides, burnt at edges, rough and impossible to colour. Inside well covered (2½″ deep).
Nov. 6	9·0	4	5	Brass 12	0	5	Grey, streaked.
Nov. 6	12·0	6	5	Brass 12	0	5	Whiter, better than last test.
Nov. 6	13·1	4·5	2·5	Brass 4	0	2	White tough plate at centre, edges badly burnt.
Nov. 6	21·4	6	5	Copper 4	0	2	Excellent deposit of good thickness, corners cracked only slightly. Deposit could be felt with finger nail; buffed to good colour. Did not crack when bent double.
Nov. 6	20·0	6·5	5	Copper 4	0	2	Same
Nov. 9	6·9	4·5	5	Copper 9	0	15	Smooth, hard, white, adherent plate. Buffed to good finish.
Nov. 11	8·3	4	5	Corrugated Steel 9·5	0	10	Best plate thus far. Buffed to mirror finish over entire surface. Deepest grooves in good condition.
Nov. 11	4·2	3·5	5	Thick Steel 28	0	50	Appeared satisfactory when removed from bath, but scaled at centre when dipped in hot water.
Nov. 12	8·6	9	5	Brass 16·5	0	1	Smooth, uniform, white plate withstood hard buffing. Coloured easily to mirror finish.
Nov. 12	8·6	8	5	Brass Hub Cap 45	0	1	Smooth, white plate, buffed to excellent finish. No evidence of a defect.
Nov. 12	19·4	8	5	Perforated Brass 8	0	2	Beautiful results; high and low spots perfect.
Nov. 12	19·4	8	5	Same Piece 8	0	1	Cleaned the piece just plated and plated on the first deposit without stripping. No indication of trouble from this source.
Nov. 12	9·7	6	5	Same	0	5	Splendid plate, coloured by rubbing with flannel cloth, afterwards stood severe buffing test.
Nov. 12	10·7	3	5	Same	0	15	Dull, muddy colour when removed from bath, buffed readily to good lustre. Background not as white as when higher voltage was used.
Nov. 17	16·8	6	5	Cast Iron 18	0	1	Burnished the piece with 400 lbs. steel balls for 15 minutes, good finish. No evidence of wearing through deposit. Placed the casting in acidulated water (15 to 1) for 36 hours, wiped dry. No evidence of defective coating, casting remained in good condition.
Nov. 17	17·1	6·5	5	Embossed Brass 12	0	1	Buffed on 10″ wheel at 3600 R.P.M., excellent finish. Did not cut through letters or raised parts.
Nov. 17	19·4	Same	5	6 of same 72	0	1	Excellent plate.
Nov. 17	26·4	Same	5	Brass 2·5	0	1	Splendid plate; hard yet pliable. Plated 200 of these pieces in same manner, in 1 doz. lots.
Nov. 17	16·9	Same	5	Polished Lead 12	0	5	Adherent plate, hard, smooth and good colour, buffed readily. Did not break when bent. Cut lead in pieces without scaling deposit.

5

54

Continued

Date of ex-periment.	Current density in amperes per sq. decimetre.	Electro-motive force in volts.	Electrode distance in inches.	Cathode area in sq. inches.	Duration of plating.		Character of plate and remarks.
Nov. 18	16·8	Same	5	Britannia Metal 30	0	5	Very satisfactory plate, adherent and white.
Nov. 18	16·8	Same	5	Steel Skate Blade 33·3	0	5	Not as white as desired, but other-wise very good plate.
Nov. 19	17·0	6	5	German Silver 10	0	2	Excellent plate.
Nov. 19	16·8	6·5	5	Tin 15	0	5	This deposit stood every conceivable abuse, and was easily coloured on small buff.
Nov. 20	16·3	Same	5	Brass Hub Cap 45	0	3	Buffed to a beautiful finish. Required very severe buffing on wheel re-volving at 1500 R.P.M. to cut through; apparently equal to a 1 hour nickel deposit.
Nov. 20	17·0	Same	5	Brass 10	0	1	Plated 100 pieces, and buffed them ready for stock in 1 hour.
Nov. 20	17·0	Same	5	Brass 60	0	1	
Nov. 23	16·8	6	5	Steel Tube 18	0	15	Drew tube from 1″ to ¾″, buffed same and found perfect, then drew tube to ⅝″ buffed again and still perfect.
Nov. 23	11·6	6	5	Oxidized Silver faced Britannia Metal 12 (14 sq. in. anode surface)	15	15	The die was merely washed with alcohol, rinsed in water and placed in cobalt bath. A current of 15 amperes was passed through the bath for 10 minutes, while the cathode was kept constantly mov-ing. Current then reduced to 9 amperes and cathode closely watched during the following hour. Bath switched to battery circuit and allowed to remain undisturbed during the night. Resulting de-posit was quite smooth at centre, edges rough but solid; face of die in fine condition. The finest lines perfect. The deposit did not curl or lift from the master die during run. Deposit about $\frac{3}{32}$″. Extraordinary results for die work.
Nov. 24	21·1	6·5	5	1/32″ Brass 4	0	5	Splendid plate, beautiful colour after buffing.
Nov. 24	16·8	6·5	2·5	Brass 12	0	5	Firm, tough, adherent plate, white in colour, began to darken at edges, oiled off edges with emery, then buffed entire surface to splendid finish.
Nov. 25	20·0	6·5	5	Brass 4	0	10	Adherent, white smooth plate, buffed very easily to good finish.

NOTE—To transform amperes per square decimetre to amperes per square foot, multiply by 9·3.

Ageing Test, Solution XIII B.

	Grams cobalt per 100 c.c. solution.	Acidity of solution.
Solution after plating as shown in above table to Oct. 6, 1914, approximately 2 hours	8·55	Strongly acid to litmus.
Solution after further run 15 hours, 15 minutes depositing approximately 37·0 grams at current density 5·35 amperes per square decimetre Oct. 7........................	8·40	Same
Solution after 17·5 hours further plating depositing about 30 grams at current density of 16·5 amperes per square decimetre Oct. 8......	8·20	Strongly acid to litmus.

Oct. 8. After all the above, a good smooth, white plate was obtained at 15·4 amperes per square decimetre in 5 minutes, which buffed readily to mirror surface.

Solution after 15 hours further plating at 11 amperes per square decimetre depositing approximately 40 grams Oct. 9.........................	8·30	Strongly acid to litmus.
Solution after 16 hours further plating at 5·0 amperes per square decimetre Oct. 10.................	7·95	Same.
Solution after 67 hours further plating at 5·26 amperes per square decimetre Oct. 13.................	7·59	Same.

Oct. 13. After all the above a good, smooth, white plate was obtained at 16·5 amperes per square decimetre in 5 minutes, which buffed readily to mirror surface.

These ageing tests were run at very severe anode current densities, so that the solution behaved remarkably well under the circumstances.

Current Efficiency Test, Solution XIII B.

Run I.

Current density approx......................	1·2 amp. per sq. dec.
Cathode—brass polished both sides—area.......	35·0 sq. cms.
Time of run................................	1 hr. 0 min.
Average current through bath................	0·435 amp.
Theoretical weight of cobalt deposited........	0·478 grams.
Weight of cathode before plating.............	30·9971 "
Weight of cathode after plating..............	31·4783 "
Cobalt deposited.......................	0·4812 "

$$\text{CURRENT EFFICIENCY} \quad \frac{0·481}{0·478} = 100·0\%$$

Run II.

Current density approx......................... 1·0 amp. per sq. dec.
Cathode—brass polished both sides—area...... 32·4 sq. cms.
Time of run.................................. 1 hr. 0 min.
Average current through bath.................. 0·324 amp.
Theoretical weight of cobalt deposited......... 0·356 grams.
Weight of cathode before plating............. 23·2262 "
Weight of cathode after plating.............. 23·5825 "

Cobalt deposited......................... 0·3563 "

$$\text{CURRENT EFFICIENCY} \quad \frac{0·356}{0·356} \quad = \quad 100·0\%$$

Run III.

Current density approx......................... 5·0 amp. per sq. dec.
Cathode—brass polished both sides—area...... 18·2 sq. cms.
Time of run.................................. 1 hr. 0 min.
Average current through bath.................. 0·922 amp.
Theoretical weight of cobalt deposited......... 1·012 grams.
Weight of cathode before plating............. 27·9628 "
Weight of cathode after plating.............. 28·9726 "

Cobalt deposited......................... 1·009 "

$$\text{CURRENT FFFICIENCY} \quad \frac{1·009}{1·012} \quad = \quad 99·6$$

SOLUTION XIII C.

Nickel sulphate $NiSO_4$................... 312·5
Sodium chloride $NaCl$................... 19·6
Boric acid H_3BO_3...................... nearly to saturation
Water................................ 1000 c.c.
This bath is the nickel analogue of XIII B.

Data of Plating Experiments, Solution XIII C.

Date of experiment.	Current density in amperes per sq. decimetre.	Electromotive force in volts.	Electrode distance in cms.	Cathode area in sq. cms.	Duration of plating.		Character of plate and remarks.
Oct. 15	3·95	5·3	10	Brass 33·8	0	20	Good, smooth, uniform deposit, nearly polished as removed from solution.
Oct. 15	5·9	3·7	10	Brass 13·5	0	15	Smooth, white plate, but splitting at edges.
Oct. 15	7·76		10	Brass 12·5	0	10	Same

Conclusions

(1) The extreme importance of proper concentration of cobalt sulphate solutions is shown by the results of this series. Solution XIII A is unsatisfactory for plating purposes at all current densities tried. Solution XIII B, which is a more concentrated solution of the same type, is the most completely satisfactory solution, for a great variety of purposes, which we have found. We know of no solution, plating with nickel, which begins to compare with solution XIII B for the range of work which it will do, and for the extreme high current densities at which it will operate. It is possible to get a plate in five minutes or less, with solution XIII B which will stand bending tests, and which will buff as satisfactorily as a plate which has taken one hour from the usual nickel plating baths.

(2) Cobalt plates from this simple cobalt sulphate solution in the presence of sodium chloride and boric acid, solution XIII B, on brass and iron, are firm, adherent, hard and uniform, and may readily be buffed to a satisfactorily finished surface. They take a very high polish, with a beautiful lustre, which, although brilliantly white, possesses a slightly bluish cast.

(3) The specific electrical conductivity of solution XIII B is much higher than that of the corresponding nickel solution.

(4) Solution XIII B does not yield the best cobalt plates at low current densities, that is, in the neighbourhood of $0\cdot50$ to $1\cdot0$ amperes per square decimetre, which is a common range for nickel plating work. Solution XIII B begins to plate most satisfactorily at a current density in the neighbourhood of $3\cdot5$ amperes per square decimetre, and continues to give satisfactory plates at all current densities up to $1\cdot75$ amperes per square decimetre. This is equivalent to a current density of more than 160 amperes per square foot, and even at this speed, the limit of the solution has not yet been reached. Later commercial tests (see p. 66) show that even higher speeds than this may be employed, under the conditions of plating practice.

(5) All of these cobalt plates within the wide current density range described as satisfactory for solution XIII B, are as smooth, adhesive and generally satisfactory as the best nickel plates.

(6) Solution XIII B does not change appreciably in cobalt content or in acidity when used over long periods of time at current densities as high as 1 ampere per square decimetre. It only showed a very gradual diminution in cobalt content under the most severe conditions of the ageing test described above. We know of no other cobalt solution and of no nickel solution which would begin to stand up under the conditions of this ageing test.

(7) There is no nickel bath of which we are aware operating in the manner of the usual commercial plating procedure at anything like as high current density as solution XIII B. See commercial tests p. 52.

(8) Solution XIII B may be used for plating on brass, iron and steel, for which cathodes the above conclusions apply.

(9) Solution XIII B may be used to deposit a heavy cobalt plate. These plates may apparently be deposited to any desired thickness, and they are firm, adherent, massive, of extreme hardness and show no tendency to curl or split.

(10) Heavy plates may be obtained from solution XIII B to much better advantage than solution XV, which has been particularly patented for the purpose with nickel, that is heavy deposits may be obtained from solution XIII B at current densities of 5 or 6 amperes per square decimetre whereas solution XV must be operated at low current densities in the neigh-

bourhood of 0·30 amperes per square decimetre. If a current density of above 6 amperes per square decimetre is used with solution XIII B for heavy deposits, under the conditions in dimension of our baths, it was found that trees were formed on the cathode.

(11) Our experiments show that solution XIII B "throws" very satisfactorily.

(12) Among the satisfactory properties of this remarkable solution should be mentioned an extremely high current efficiency, which we found at 1·0 and 5·0 amperes per square decimetre, to be almost 100 per cent.

(13) Solution XIII C, which is the nickel analogue of solution XIII B yielded satisfactory plates up to about 5 amperes per square decimetre, but showed splitting at current densities greater than that. Nickel solution XIII C does not possess the remarkable qualities of its cobalt analogue XIII B, although in many respects it is an improvement on standard nickel solutions.

(14) Solution XIII B requires very little ageing; it operates satisfactorily almost from the start.

(15) Solution XIII B is so remarkable in its properties that it was thought highly worth while to develop it further under commercial conditions. See commercial tests pp. 52-4, and 65-67.

SERIES 14: COBALT-SULPHATE, AMMONIUM-SULPHATE, MAGNESIUM SULPHATE, BORIC ACID SOLUTIONS.

SOLUTION XIV.

Cobalt sulphate $CoSO_4$.................. 37·5 grams.
Ammonium sulphate $(NH_4)SO_4$.......... 21·7 "
Magnesium sulphate $Mg SO_4$............. 3·3 "
Boric acid $H_3 BO_3$..................... 12·1 "
Water................................. 1000 c.c.
       Total bath........................ 3·5 litres.

Data of Plating Experiments, Solution XIV.

						Cast Anodes
Date of experiment.	Current density in amperes per sq. decimetre.	Electro-motive force in volts.	Electrode distance in cms.	Cathode area in sq. cms.	Duration of plating.	Character of plate and remarks.
Oct. 1	0·50	1·05	10	Brass 32·4	2 0	Uniform, smooth white deposit, readily buffed to mirror surface.
Oct. 1	1·25	1·92	10	Brass 35·3	1 0	Same
Oct. 2	1·50	2·20	10	Brass 35·3	1 0	Same
Oct. 2	1·75	1·68	10	Brass 18·5	0 45	Same
Oct. 2	2·0	1·55	10	Brass 11·1	1 30	Same
Oct. 2	2·5	1·77	10	Brass 11·3	1 30	Same
Oct. 3	3·0	2·47	10	Brass 17·2	0 30	Same
Oct. 5	3·0	3·32	10	Brass 19·6	0 20	Uniform, smooth white deposit, slightly burned on edges.
Oct. 3	3·51	2·88	10	Brass 16·8	0 30	Uniform smooth white deposit, readily buffed to mirror surface.
Oct. 5	4·0	2·98	10	Brass 13·3	0 20	Uniform, smooth white deposit, slightly burned on edges.

Current Efficiency Test, Solution XIV.

Run I.

Current density approx......................	1·0 amp. per sq. dec.
Cathode—brass polished both sides—area......	51·0 sq. cms.
Time of run.................................	1 hr. 0 min.
Average current through bath................	0·513 amp.
Theoretical weight of cobalt deposited........	0·564 grams.
Weight of cathode before plating............	61·8019 "
Weight of cathode after plating..............	62·3503 "
Cobalt deposited...........................	0·5484

$$\text{CURRENT EFFICIENCY} \quad \frac{0\cdot548}{0\cdot564} = 97\cdot2\%$$

Run II.

Current density approx......................	1·0 amp. per sq. dec.
Cathode—brass polished both sides...........	40·0 sq. cms.
Time of run.................................	1 hr. 0 min.
Average current through bath................	0·402 amp.
Theoretical weight of cobalt deposited........	0·442 grams.
Weight of cathode before plating............	57·6491 "
Weight of cathode after plating..............	58·0746 "
Cobalt deposited...........................	0·4255

$$\text{CURRENT EFFICIENCY} \quad \frac{0\cdot426}{0\cdot442} = 96\cdot3\%$$

CONCLUSIONS.

(1) Cobalt plates from solution XIV, on brass and iron, are firm, adherent, hard and uniform, and may readily be buffed to a satisfactory mirror surface. They take a very high polish with a beautiful lustre, which, although brilliantly white, possesses a slightly bluish cast.

(2) Solution XIV yields satisfactory plates at all current densities up to about 3 amperes per square decimetre.

(3) The current efficiency of solution XIV is satisfactorily high, being in the neighbourhood of 96 to 97 per cent under the conditions of our experiments.

(4) All of the cobalt plates from solution XIV within the current density range described as satisfactory, are as smooth, adhesive and generally satisfactory as the best nickel plates.

SERIES 15: COBALT ETHYL SULPHATE SODIUM SULPHATE AMMONIUM CHLORIDE SOLUTIONS.

SOLUTION XV.

This solution is the cobalt analogue of a nickel solution patented by Dr. G. Langbein and Co., Leipzig, Germany.[1] The inventor claims for this solution that very dense, hard, uniform, deposits of nickel may be obtained from it, and particularly "deposits of any desired thickness can be

---

[1] Kaiserliches Patenamt, Patentschrift No. 134736 K1 48a Sept. 18th, 1902.

produced if the bath be constantly agitated by mechanical means or by the introduction of hydrogen." It, of course, is not permissible to agitate this bath with air as that would cause oxidation.

Cobalt ethyl sulphate..................	100 grams.
Sodium sulphate......................	10·0 "
Ammonium chloride....................	5·0 "
Water...............................	1000 c.c.
Total bath.......................	2 litres.

We made a number of cobalt depositions with solution XV at various current densities, both with and without mechanical agitation and for varying lengths of time. As a result of these experiments it was found that mechanical agitation with cobalt bath was not necessary as in the case of the nickel bath, indeed it always caused the plate to crack.

One of the depositions was continued for a period of 8 twenty-four hour days, at the end of which time a uniform dense, hard, satisfactory plate was found. The thickness of this plate was approximately half a millimetre. It was deposited at a current density of 0·30 amperes per square decimetre. This plate was, of course, not as hard a deposit at this latter current density as some of the heavy plates from solution XIII B, which were deposited at very much higher current densities.

CONCLUSIONS.

(1) Dense, hard, uniform deposits of cobalt may be obtained from a cobalt ethyl sulphate solution made up like solution XV, without mechanical agitation, provided that the current density be low, not exceeding 0·30 amperes per square decimetre.

(2) For heavy depositions of cobalt, where density, hardness and speed of deposition are important, solution XV is not nearly as satisfactory as solution XIII B. However, the cobalt ethyl sulphate solution deposits cobalt more satisfactorily than the corresponding nickel ethyl sulphate solution deposits nickel.

SERIES 16: COBALT SULPHATE, AMMONIUM SULPHATE, AMMONIUM CHLORIDE, BORIC ACID SOLUTIONS.

SOLUTION XVI A.

This bath was made up as follows:—

Cobalt sulphate $CoSO_4$..................	90·7 grams.
Ammonium sulphate $(NH_4)_2SO_4$..........	27·6 "
Ammonium chloride NH_4Cl..............	15·0 "
Boric acid H_3BO_3.....................	5·2
Water................................	1000 c.c.

Bath was neutral before using, by slight addition of ammonia.

Solution XVI A and its analogue XVI B, which was identical with it, except that nickel sulphate was substituted for cobalt sulphate, were both used for an extended series of plating experiments. Both were found to give satisfactory plates at low current densities, but to fail by splitting and burning when current densities of 1·0 ampere per square decimetre or more were reached.

In making this solution it was reasoned, as in many instances already reported, that a greater current density might be used if the metal content were increased. This solution, contrary to a number of others reported, would not, however, admit of very great increase in cobalt content without giving disturbing effects of crystallization in the bath.

A solution containing 150 grams cobalt ammonium sulphate $CoSO_4 \cdot (NH_4)_2SO_4$, with an excess of $11 \cdot 3$ grams cobalt sulphate $CoSO_4$, was found to precipitate a red compound on the anode.

Solution XVI A and its nickel analogue were run simultaneously in series, and a very much greater hardness of the cobalt plate than the nickel plate was particularly noticeable throughout.

Current efficiency runs were made with these solutions, which show them to be in the neighbourhood of 90 per cent, but they were not thought of sufficient importance to warrant a careful study.

CONCLUSIONS.

(1) Plates from solution XVI A and the corresponding nickel solution are both satisfactory at current densities below 1 ampere per square decimetre. At higher current densities both solutions fail.

(2) A solution of the type of Series 16 cannot be prepared, much greater in concentration than XVI A, without obtaining troublesome crystallization. Therefore, Series 16 does not offer a highly concentrated rapid plating cobalt solution.

(3) Cobalt plates from solution XVI A, on brass and iron, are firm, adherent, hard and uniform, and may readily be buffed to a satisfactory surface, within current density range recommended. They take a very high polish, with a beautiful lustre, which, although brilliantly white, possesses a slightly bluish cast.

(4) The cobalt plates from solution XVI A are very much harder than the nickel plates from its analogue XVI B. This is noticeable throughout all our plating experiments, but is particularly striking in this case where the plates were obtained in series and simultaneously.

(5) The specific electrical conductivity of solution XVI A is very much higher than that of the corresponding nickel solution XVI B.

(6) The current efficiencies of both nickel and cobalt solutions of the type XVI A are in the neighbourhood of 90 per cent under the conditions of our experiments.

DECOMPOSITION VOLTAGES AND POLARIZATION.

Throughout this paper, wherever we have found that a given cobalt solution plated at a higher current density than the corresponding nickel solution, at the same electromotive force, we have attributed it to the higher specific electrical conductivity of the cobalt solution. This has been true for nearly all the solutions which we have studied, for we have found that with a cobalt solution a lower voltage yields a higher current density than will the corresponding nickel solution.

It is, of course, not necessary that the cause of this condition should be a higher conductivity of the cobalt solution, for it might be that there was an electromotive force of polarization in the circuit, under the conditions of these practical plating experiments, which were higher for the nickel solutions than for the corresponding cobalt solutions.

Although we had no reason to suspect any irregularity in our experiments, and while normally the electromotive force of polarization with the cobalt or nickel solutions, used with a soluble cobalt or nickel anode, and plating upon cobalt or nickel, would be negligibly small. After the first few moments nevertheless we ran a few simple experiments to cover these points.

While we were making the tests on the cells, cobalt-electrolyte-cobalt, and nickel-electrolyte-nickel, under the conditions of our plating experiments, it was thought worth while to measure the relative decomposition voltages of normal cobalt sulphate and normal nickel sulphate solutions.

Method of Measurement.

For these measurements the cell was placed in series with an appropriate regulating resistance and a milliammeter, across storage batteries capable of giving electromotive forces up to 8 volts. A voltmeter was connected across the electrodes of the cell. A carefully made switch, conveniently arranged for rapid and accurate operation, was placed in the main circuit, and a key was similarly placed in the voltmeter circuit. The method of reading the polarization consisted in opening the line switch with one hand, thereupon closing the switch to the voltmeter circuit with the other. There was, of course, an appreciable time interval between the opening of the first switch and the closing of the second, so that the actual value of polarization must be slightly higher than the values recorded. Moreover the voltmeter requires a certain amount of current, however small, for its operation, which also caused the recorded values to be slightly lower than the real values.

However, the magnitude of the voltages recorded in these measurements is so large that the error, in most instances, is relatively small, and moreover the primary purpose at present is to compare the polarization values for nickel sulphate and cobalt sulphate.

Data—Decomposition Voltages and Polarization.

$CoSO_4$ and $NiSO_4$.

The diagrams (Figs. I to IV) following, are plots of the measurements of impressed voltages, current and voltages of polarization, as determined by method of measurment described above. In each instance a second complete set of figures was obtained, agreeing closely with those shown.

The platinum electrodes used for these polarization experiments were $2''$ in depth and $1''$ in width.

From the plots, Figs. I—IV, the following conclusions may be drawn:—

CONCLUSIONS.

(1) Under the conditions of our experiments the decomposition pressure of normal cobalt sulphate is approximately $2 \cdot 16$ volts, whereas correspondingly that of normal nickel sulphate is approximately $2 \cdot 41$ volts.

(2) The polarization for both cobalt and nickel solutions, under the conditions of the plating bath, where both anode and cathode are respectively of cobalt or nickel, are negligibly small, and not very different from one another in magnitude.

(3) The advantage which we have found throughout these researches for cobalt solutions over the corresponding nickel solutions, in that a much smaller impressed electromotive force will yield a given current density with cobalt than with nickel, is due to the greater specific electrical conductivity of the cobalt solutions, and not to some effect of polarization.

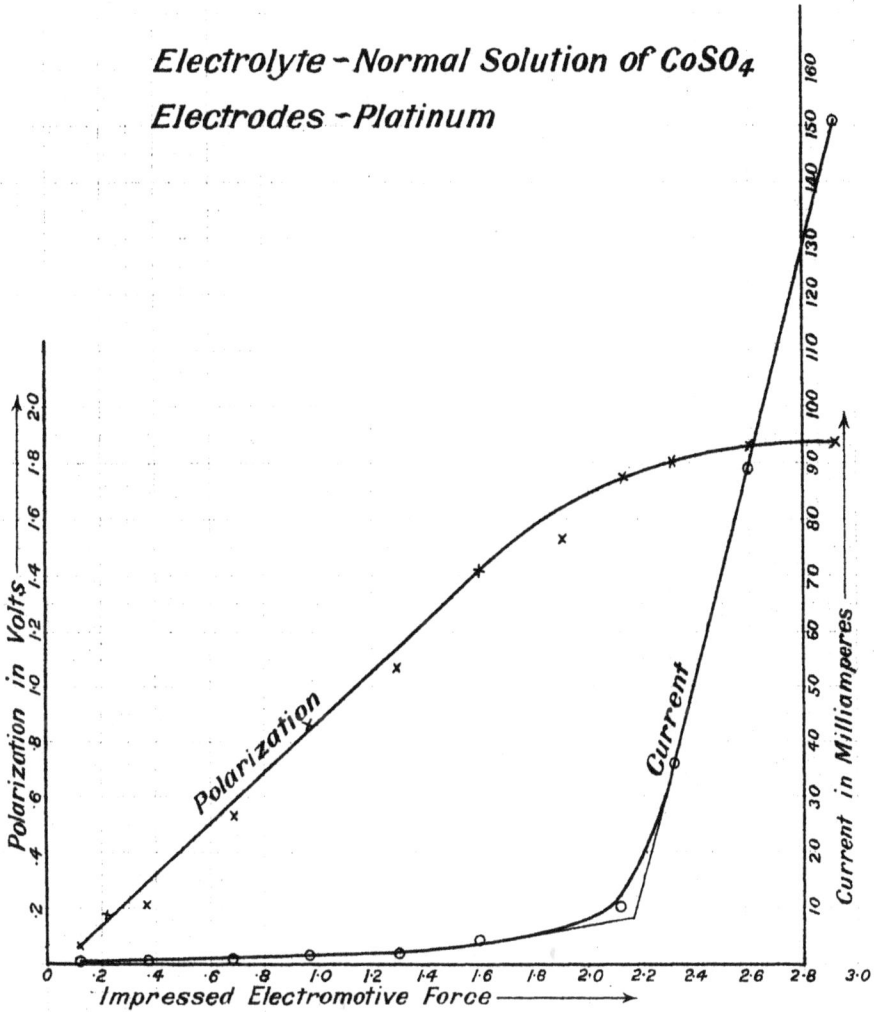

Figure I

Electrolyte – Normal Solution of $CoSO_4$
Electrodes – Platinum

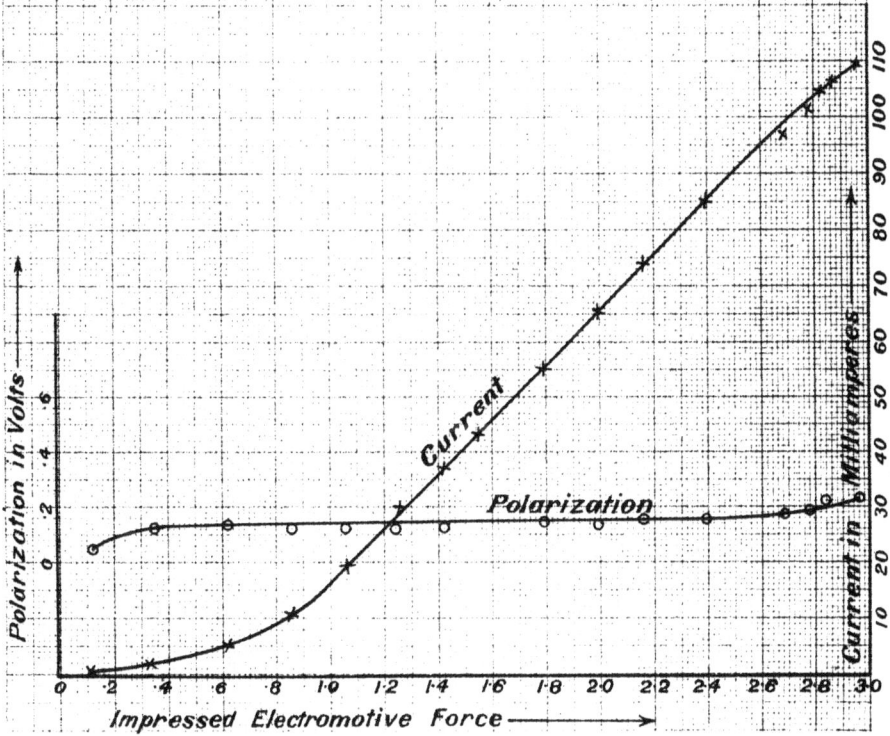

Figure II

Electrolyte - Normal Solution of CoSO₄
Electrodes - Cobalt

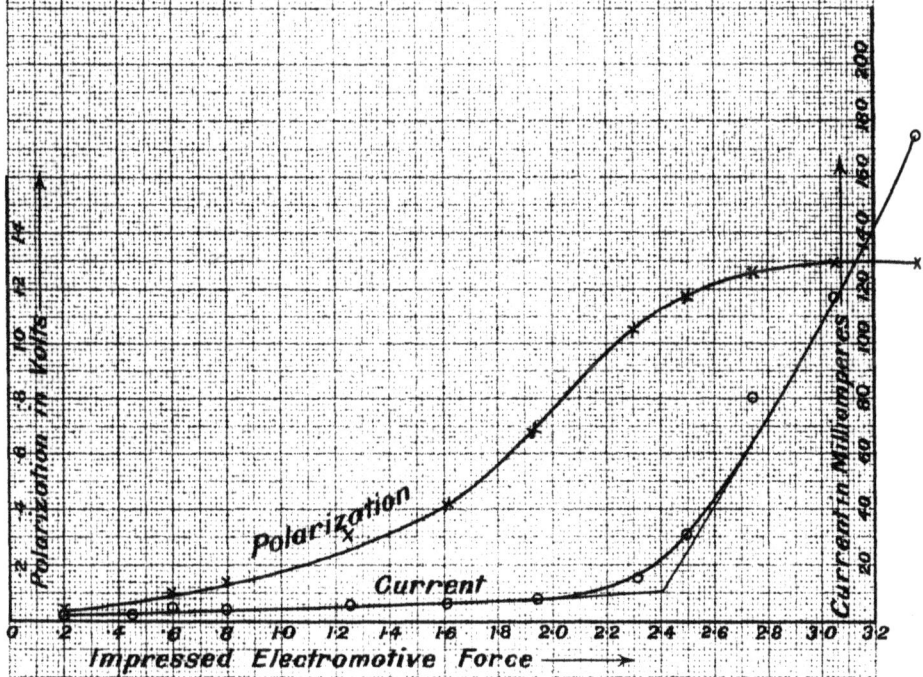

Figure III

Electrolyte - Normal NiSO₄
Electrodes - Platinum

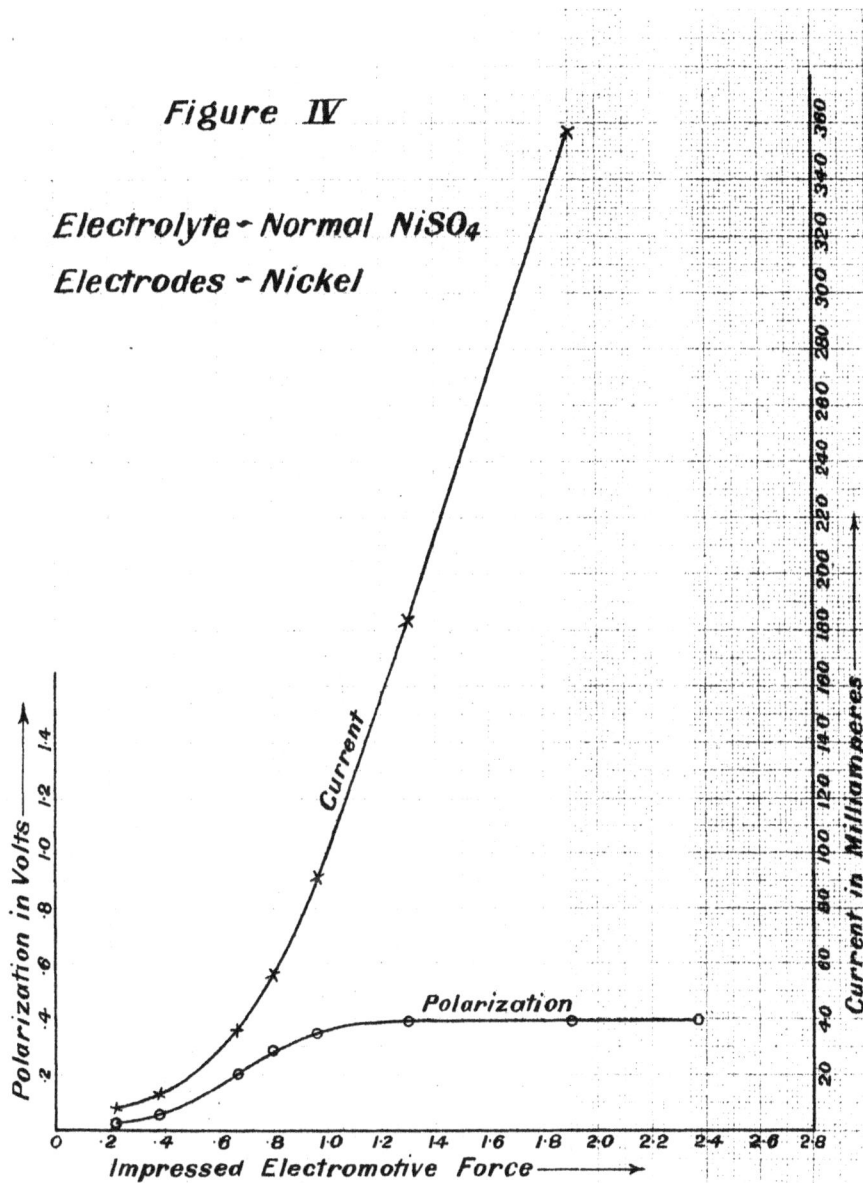

Figure IV

Electrolyte - Normal NiSO₄

Electrodes - Nickel

COMMERCIAL TESTS WITH SOLUTIONS I B AND XIII B.

Verification of Plating Results on Commercial Scale.

The results with solution I B (and later with solution XIII B) were thought by the authors to be so unusual and of sufficient importance to warrant verification and further development under standard commercial conditions. We, therefore, arranged with the Russell Motor Car Company of Toronto, Ontario, and particularly with Mr. W. S. Barrows, foreman electroplater, with the same Company, to have a plating tank operated under standard commercial conditions at their plant, using solution I B and later solution XIII B.

Mr. Barrows has had some twenty years of experience with all sorts of electro-plating work, to which he has particularly devoted himself. He has considered these two solutions entirely from the point of view of commercial practicability and value.

The authors take great pleasure in expressing their thanks to the Russell Motor Car Company, and to Mr. W. S. Barrows, for their collaboration, and particularly to Mr. Barrows for the careful, vigorous and painstaking manner in which he has subjected the solutions and plates to the various tests that were required, to establish their commercial importance and value.

SOLUTION I B.

Salts identical with those used in this laboratory were sent by the authors to Mr. Barrows, with instructions for making up solution I B identical with that used by us and described on page 14.

Anodes.

Cobalt anodes were cast at this laboratory of the size required by Mr. Barrows for his tank, and sent him for use in these experiments. They analysed as follows:—

Co.	98·75
Ni.	none
Fe.	1·35
As.	none
P.	0·0067
S.	0·052
C.	0·061

The anodes used in a bath had a surface of approximately 1 square foot area.

Commercial Tests at Russell Motor Car Co.

Report of Mr. W. S. Barrows, foreman electroplater.

Mr. Barrows, in collaboration with the authors, tested solution I B for plating purposes, during the months of August and September, 1914. The plating was accomplished under standard commercial conditions on copper, brass, iron, steel, tin, German silver, lead, and Britannia metal. Various articles, such as brass castings, sheet brass, steel stampings, skates, automobile hubs, etc., etc.,—articles of very different shapes and sizes, were plated under exactly the same general conditions as for nickel plating practice at the Russell Motor Car Company.

The tests were made in a still solution, that is without agitation of any kind, and the resulting plates were subjected to the most severe practical tests. This work was regarded by Mr. Barrows purely from the commercial view point, and with this in mind, he particularly tested and studied the following points—the colour of plate, the uniformity and freedom from defects of the plates, the allowable speed of plating without pitting or burning, or the maximum allowable current density, the solubility of the anodes, the required voltage, the "throwing" properties of the bath, that is, its ability to cover the deeper parts of the object in a satisfactory manner, the solubility of the salts, the hardness of the plates, the ease of "colouring" the plate on a buff, the efficiency of the plating solution, the time required for ageing of the bath, the adhesiveness of the plate to the cathode under bending and hammering tests, the general cleanliness of the bath, the corrosion of the plate, besides many other special features.

In a letter to one of the authors, dated November 2, 1914, Mr. Barrows gives a very complete report of his commercial tests of solution I B. This letter follows in full as received, and serves admirably to cover this portion of the work.

628 Dovercourt Rd.,
Toronto, Nov. 2nd, 1914.

Dr. Herbert T. Kalmus,
    Queen's University,
      Kingston, Ont.

Dear Sir,—

After preparing a cobalt plating solution according to your formula for bath I B, and having used this bath daily during the past eight weeks, plating a great variety of copper, brass, iron, steel, tin, German silver, lead and Britannia metal articles of different shapes and sizes under exactly the same conditions as met with in general nickel plating at the factory of the Russell Motor Car Company, West Toronto, and after regarding the characteristics of this particular solution absolutely from a commercial viewpoint, I can heartily confirm any statement you have made to me regarding this remarkable solution. This bath was equipped with cobalt anodes, 98·75 per cent cobalt, which were sent to me from your laboratory.

The runs made have varied from five minutes, to 24 hours, and in each case the bath has proved wonderfully efficient.

The cobalt plates obtained were smooth, white and fine grained, very adherent and uniform. In fact the surfaces of these deposits after several hours run were so very smooth and uniform, that a 4 inch cotton buff colored them to a mirror finish quite easily. We use 14 inch and 16 inch buffs to color 3 hour deposits of nickel.

To test the hardness of the cobalt as compared with nickel, with reference to either buffing or polishing with emery, we plated strips of brass, one half the surface with cobalt and one half with nickel, always giving the nickelled portion the thickest plate. Then buffing or polishing across the two deposits we found invariably that the nickel was removed from the brass before the cobalt, and in some cases in one half the time.

Though so hard and firm, these plates color beautifully with little effort, and require the use of much less buffing composition than comparatively thin plates of nickel. Automobile parts of irregular shape were plated from 10 to 20 minutes, and finished on a 6 inch buff operated at 3,000 R.P.M. without the slightest evidence of a defect in the plating. To accomplish this with our fastest nickel baths would require at least 60 minutes of plating.

As a protective coating for iron or steel surfaces I am convinced that a comparatively thin plate of cobalt will prove equally as effective as a thick plate of nickel from an ordinary double sulphate nickel bath, and the time and power required for the production of such plates is decidedly in favor of the cobalt.

The deposits are also very adherent, no difficulty having been experienced in this respect, although tests were made repeatedly by bending, hammering and burnishing.

One of the weak points of several so-called rapid nickel plating solutions which we have tried commercially, is their poor "throwing" powers, - - i.e. they do not deposit the nickel readily in the indentations or cavities of the cathode. The cobalt solution I B, meets this requirement in a most efficient manner, the deposits on the distant portions of the cathode withstand the tests imposed in every case.

Another most important feature of this solution, which should commend itself to every practical plater and manufacturer of plated wares, is the extremely high current density at which this solution may be employed without danger of pitting the plated surface. I have plated with this cobalt solution I B, satisfactorily and under commercial conditions, at a current density of 42 amperes per square foot. This is $4\frac{1}{5}$ times the speed of our fastest commercial nickel solutions.

As a further test we plated steel tubes of 1″ diameter for two hours, with a current density of 27 amps. per square foot, and then drew the tubes down to $\frac{5}{8}$″ diameter without injuring the deposit. Though extremely hard, the ductility of the deposited metal proved remarkable.

All of our tests have been made in a still solution, without agitation of any kind, and the plates were subjected to the most severe treatment considered practical for high grade metallic coatings on the various metals heretofore mentioned.

We are also of the opinion that the anodes in the cobalt bath I B will remain free from coatings, such as characterize average anodes used in nickel baths, and that the cost of maintenance will be practically nothing compared to double sulphate nickel solutions.

I can assure you that my experience thus far with these cobalt solutions has been intensely interesting, and I sincerely believe that their use commercially would revolutionize the art of electro plating such wares as are now nickel plated.

The simplicity of its composition, its self sustaining qualities, the remarkable speed of deposition, together with the several points mentioned previously, should appeal to the commercial requirements of this progressive age. I remain.

Very truly yours,
(*Signed*) **Walter S. Barrows,**
Foreman Electroplater,
Russell Motor Car Co.,
West Toronto, Ont.

SOLUTION XIII B.

After the completion of the tests on solution I B by Mr. Barrows, salts identical with those used by the authors were sent to him, with instructions for preparing solution XIII B, identical with that described on p. 51.

Anodes. Identical with those used for solution I B, see p. 63.

Commercial Tests at Russel Motor Car Co.

Report of Mr. W. S. Barrows, foreman electroplater.

Tests were made of this solution in the manner and from the same point of view as those for solution I B (see pp. 63-64).

In a letter to one of the authors, dated December 1, 1914, Mr. Barrows gives a complete report of his commercial tests of solution XIII B. This letter follows in full, as received:—

628 Dovercourt Rd.,
Toronto, Dec. 1, 1914.

Dr. Herbert T. Kalmus,
Queen's University,
Kingston, Ont.

Dear Sir,—

After thoroughly testing cobalt plating bath XIII B, made according to your formula, I take pleasure in submitting the following report.

I found the bath very simple to prepare and at once began to operate the solution with high current densities. The results obtained were exceedingly gratifying. Evidently bath XIII B will require no prolonged ageing treatment, as splendid, white, hard, perfect deposits were obtained with extremely high current densities within three hours after bath was prepared.

The experiments have been varied and the tests of plates severe and deliberate, the results have invariably been such as to cause me to regard cobalt bath XIII B the greatest achievement in modern electro plating improvements.

The operation of the bath is positively fascinating, the limit of speed for commercial plating is astonishing, while the excellence of the plates produced is superior to those of nickel for many reasons.

The efficiency of the freshly prepared solution, together with the self sustaining qualities of the bath are without a parallel in any plating solution of any kind I have ever used.

Thin embossed brass stampings were plated in bath XIII B for only one minute, then given to a buffer who did not know the bath existed and who was accustomed to buffing $1\frac{1}{4}$ hour nickel deposits on these same stampings. This man buffed the cobalt plates upon a 10″ cotton buff wheel revolving at 3000 R.P.M. The finish was perfect with no edges exposed. These stampings have been plated in two dozen lots for one minute and from a total of 500 stampings we have found but three stampings imperfect after buffing. Each stamping is formed to a spiral after finishing without injury to the deposit. Grey iron castings with raised designs upon the surface were plated one minute in cobalt bath XIII B then burnished with 400 lbs. of $\frac{1}{8}$″ steel balls for $\frac{1}{4}$ hour without the slightest injury to the cobalt coating, as was proven by a 36 hour immersion in 15 ozs. of water acidulated with 1 oz. of sulphuric acid.

While attempting to reach the limit of current densities which would be practical with this bath XIII B, I have plated brass automobile trimmings with a current density of 244 amperes per square foot. These pieces were plated in lots of 6, and a total of 100 were plated, buffed and ready for stock in 1 hours' time. No unusual preparation was made for the run and the work was performed by one man. Size of piece plated $1\frac{1}{2}$″ × 5″.

Automobile hub caps were plated three minutes in cobalt bath XIII B and buffed to a beautiful lustre of deep rich bluish tone by use of a 7″ cotton buff revolving at 1200 R.P.M. The deposits were ample for severe treatment usually received by such articles. Comparative tests of these deposits were made as follows. Same style castings plated in double sulphate nickel solution one hour were suspended as anodes in a solution of equal parts muriatic acid and water, sheet lead cathodes were used and a current of 200 amperes at 10 volts passed through the bath. The nickel was removed from the castings in 30 seconds while 45 seconds time was required to remove the cobalt plates.

The above mentioned plating tests were made with still solution, no form of agitation being employed. By aid of mechanical agitators these current densities could be greatly exceeded with highly satisfactory results.

These cobalt plates were very hard, white and adherent and colored easily with slight effort.

Several plates were produced upon sharp steel surgical instruments, these instruments finished perfectly and owing to the hardness of the cobalt plate only a thin deposit was required to equal the best nickel deposits which we received as samples. Cobalt deposits should prove especially valuable for electro plating surgical instruments for this reason, nonadherent thick deposits being very dangerous for this class of work.

Owing to the unusual mild weather in this locality during the past month, I have not concluded test with cobalt plates on highly tempered nickel steel skate blades, but judging from appearances and various severe indoor tests we do not hesitate to report success in this direction. A three minute deposit from bath XIII B resists corrosion equally as long as a one hour nickel deposit, the finish is even superior to nickel, while every test employed during the process of manufacturing the nickel plated article has proven equally ineffective with cobalt plates, therefore by reason of the effectiveness of thin cobalt deposits we believe cobalt plates should prove wonderfullyefficient on skates, or any keen edged tool requiring a protective metallic coating.

The runs made with bath XIII B have varied from one minute to $15\frac{1}{2}$ hours, and in each case the results were remarkable. Electrotypes were reproduced 1/16″ thick. Electro dies were faced with cobalt $\frac{1}{8}$″ thick, the electrotypes being graphite covered wax and lead moulds, while the dies were made on oxidized silver faced Britannia metal.

The deposits from cobalt bath XIII B were very adherent and pliable; by proper regulation of the current beautiful white, hard, tough plates may be produced quickly on any conducting surface.

The "throwing" powers of cobalt bath XIII B make possible its employment for plating deeply indented or grooved articles such as reflectors, channel bars or articles with projecting portions.

We also obtained the best plates with extremely high current densities, although plates finished with 75 amperes per square foot were of good color and easily buffed. The production of excellent plates with a current density of 150 amperes proved particularly easy and densities in this neighborhood were employed for the greater portion of our tests.

Cobalt bath XIII B will produce excellent hard, white, tough plates absolutely free from pits or blemish at a current density of 150 amperes per square foot and under ordinary commercial conditions. This is 15 times the speed of our fastest commercial nickel solution.

Furthermore, the anode tops and hooks remain free from creeping salts. The solution retains its original clean appearance and the anodes dissolve satisfactorily, no slime or coating formed, brushing or cleaning anodes therefore will be unnecessary. The anodes used with this bath were 98·75 per cent cobalt which were sent me from your laboratory. The bath at the commencement of our tests was strongly acid to litmus, and has remained unchanged throughout our experiments. The specific gravity of the solution when freshly prepared was 1·24 and is the same to-day.

The rich deep bluish white tone of the cobalt plates upon polished brass surfaces is particularly noteworthy, this feature should assist greatly in making cobalt deposits very popular for brass fixtures, trimmings and plumbers, supplies.

My experience with cobalt bath XIII B is by no means at an end. I intend to continue its use until present supplies are exhausted and then equip a larger bath if supplies are obtainable. As a commercial proposition I am satisfied it is wonderfully efficient and economical.

Taking into account the difference in cost of cobalt as compared with nickel, I am satisfied the metal costs for plating a given quantity of work with cobalt, would be considerably less than for nickel plating a like quantity.

Furthermore the use of cobalt bath XIII B equipped with automatic apparatus for conveying parts through the bath would reduce the labor cost 75 per cent, such apparatus would be practical for a greater variety of wares than is now the case with nickel.

We cannot speak too highly of cobalt bath XIII B, and confidently believe its future history will surpass the history of any electro plating bath now in general use.

In conclusion, please accept my warmest congratulations upon your successes with cobalt solutions, and heartily appreciating the opportunity of testing these solutions, I desire to sincerely thank you, kind sir, for the benefits derived therefrom.

Very truly yours,

(*Signed*) **Walter S. Barrows,**

Foreman Electroplater,
Russell Motor Car Co.,
West Toronto, Ont.

GENERAL CONCLUSIONS FROM COMMERCIAL TESTS ON COBALT PLATING SOLUTIONS.

(1) Several cobalt solutions were found to be suitable for electroplating with cobalt under the conditions of commercial practice. Best among these are the following:—

SOLUTION I B.

Cobalt-ammonium-sulphate, $CoSO_4 \cdot (NH_4)_2SO_4 \cdot 6H_2O$, 200 grams to the litre of water, which is the equivalent of 145 grams of anhydrous cobalt-ammonium-sulphate, $CoSO_4 \cdot (NH_4)_2SO_4$, to the litre of water. Sp. gr. $= 1 \cdot 053$ at 15°C.

SOLUTION XIII B.

Cobalt sulphate $CoSO_4$.................. 312 grams.
Sodium chloride NaCl.................. 19·6 "
Boric acid........................... nearly to saturation
Water................................ 1000 c.c.
        Sp. Gr. $= 1 \cdot 25$ at 15°C.

(2) Cobalt plates from these solutions, on brass, iron, steel, copper, tin, German silver, lead and Britannia metal articles, of different shapes and sizes, deposited under conditions identical with those met with in general nickel plating practice, are firm, adherent, hard, and uniform. They may readily be buffed to a satisfactorily finished surface, having a beautiful lustre, which, although brilliantly white, possesses a slightly bluish cast.

(3) The electrical conductivity of these solutions is considerably higher than that of the standard commercial nickel solutions, so that other things being equal, they may be operated at a lower voltage for a given speed of plating.

(4) Solution I B is capable of cobalt plating on the various sizes and shapes of objects met with in commercial practice at a speed at least four times that of the fastest satisfactory nickel solutions.

(5)   Solution XIII B is capable of cobalt plating on the various sizes and shapes of objects met with in commercial practice at a speed at least fifteen times as great as that of the fastest satisfactory nickel solutions.

(6)   Plates from both of these solutions on various stock pieces, satisfactorily withstood the various bending, hammering and burnishing tests to which commercial nickel work is ordinarily submitted.

(7)   These two very rapid cobalt solutions are remarkable for their satisfactory throwing power. That is, they readily and satisfactorily deposit the cobalt in the indentations of the work.

(8)   These two rapid solutions operate at these high speeds in a perfectly still solution without agitation of any kind.

(9)   These solutions are both cleaner, that is free from creeping salts and precipitated matter, than the standard commercial nickel baths.

(10)   The cobalt deposited at this rapid speed is very much harder than the nickel deposited in any commercial nickel bath. Consequently a lesser weight of this hard cobalt deposit will offer the same protective coat as a greater weight of the softer nickel deposit. Considering solution XIII B, operating at 150 amperes per square foot, on automobile parts, brass stampings, etc., etc., a sufficient weight of cobalt to stand the usual commercial tests, including buffing and finishing, is deposited in one minute. With the best nickel baths, it takes one hour, at about 10 amperes per square foot, to deposit a plate equally satisfactory. Therefore, the actual weight of metal on the cobalt plate must be approximately one quarter that of the nickel.

(11)   For many purposes, under the condition of these rapid plating solutions, one-fourth the weight of cobalt, as compared with nickel, is required to do the same protective work. Consequently, if nickel is worth 50c a lb., in the anode form, cobalt could be worth nearly $2 a lb., in the same form, to be on the same basis, weight for weight of metal. In addition there are other advantages of cobalt in saving of labour, time, overhead, etc.

(12)   A smaller plating room would handle a given amount of work per day with cobalt than with nickel.

(13)   With these very rapid plating solutions, by the use of mechanical devices to handle the work, the time required for plating, as well as the labour costs may be tremendously reduced. Solution I B, and particularly Solution XIII B, are so rapid as to be revolutionary in this respect.

(14)   Obviously the cost of supplies, repairs, etc., would be less with cobalt plating than with nickel plating, as the size of the plant for a required amount of work is less.

(15)   The voltage required for extremely rapid cobalt plating is greater than that for most nickel plating baths; it is not so great but that the machines at present in use may in general be operated. For the same speed of plating, the cobalt solution requires much the lower voltage.

(16)   For a given amount of work the power consumption for this rapid cobalt work is less than that for nickel. This is obvious, because the total amount of metal deposited in the case of cobalt is very much less, whereas the voltage at which it is deposited is not correspondingly greater.

(17)   Ornamental work on brass, copper, tin, or German silver would require only a one minute deposit. Even wares exposed to severe atmospheric influences, or friction, could be admirably coated with cobalt in solution XIII B in fifteen minutes. The tremendous possibilities of this solution are not to be completely realized unless mechanical devices are applied to reduce hand labour to a considerable extent.

(18) Thick deposits from these solutions are vastly superior to any that we have seen produced from nickel solutions. The tendency to distort thin cathodes is less pronounced, while electrotypes and electro-dies have been given a superior thick deposit in a most satisfactory manner. The lines were hard, sharp and tough and the surface smooth. Nickel does not equal cobalt for excellence of massive plates.

(19) Many of these tests were passed upon by uninterested skilled mechanics at the plant of the Russell Motor Car Company, who invariably reported in favour of cobalt as above.

(20) Both solutions I B and XIII B are substantially self-sustaining, once they are put into operating condition, and the amount of ageing required to do this is very much less for them than that for the present commercial nickel baths.

ACKNOWLEDGMENTS.

The numerous analyses in connexion with these experiments were made by Mr. Russell C. Wilcox, B.Sc., and during the summer of 1914, a number of plating experiments were run by Mr. C. S. Allin, B.A. The authors acknowledge with thanks their indebtedness to these gentlemen.

CANADA
DEPARTMENT OF MINES

HON. LOUIS CODERRE, MINISTER; R. G. McCONNELL, DEPUTY MINISTER

MINES BRANCH

EUGENE HAANEL, PH.D., DIRECTOR.

REPORTS AND MAPS

PUBLISHED BY THE

MINES BRANCH

REPORTS.

1. Mining conditions in the Klondike, Yukon. Report on—by Eugene Haanel, Ph.d., 1902.

†2. Great landslide at Frank, Alta. Report on—by R. G. McConnell, B.A., and R. W. Brock, M.A., 1903.

†3. Investigation of the different electro-thermic processes for the smelting of iron ores and the making of steel, in operation in Europe. Report of Special Commission—by Eugene Haanel, Ph.D., 1904.

5. On the location and examination of magnetic ore deposits by magnetometric measurements—by Eugene Haanel, Ph.D., 1904.

†7. Limestones, and the lime industry of Manitoba. Preliminary report on—by J. W. Wells, M.A., 1905.

†8. Clays and shales of Manitoba: their industrial value. Preliminary report on—by J. W. Wells, M.A., 1905.

†9. Hydraulic cements (raw materials) in Manitoba: manufacture and uses of. Preliminary report on—by J. W. Wells, M.A., 1905.

†10. Mica: its occurrence, exploitation, and uses—by Fritz Cirkel, M.E., 1905. (See No. 118.)

†11. Asbestos: its occurrence, exploitation, and uses—by Fritz Cirkel, M.E., 1905. (See No. 69.)

†12. Zinc resources of British Columbia and the conditions affecting their exploitation. Report of the Commission appointed to investigate —by W. R. Ingalls, M.E., 1905.

†16. *Experiments made at Sault Ste. Marie, under Government auspices, in the smelting of Canadian iron ores by the electro-thermic process. Final report on—by Eugene Haanel, Ph.D., 1907.

†17. Mines of the silver-cobalt ores of the Cobalt district: their present and prospective output. Report on—by Eugene Haanel, Ph.D., 1907.

* A few copies of the Preliminary Report, 1906, are still available.
† Publications marked thus † are out of print.

† Publications marked thus † are out of print.

NOTE. — *The following parts were separately printed and issued in advance of the Annual Report for 1907-8.*

†31. Production of cement in Canada, 1908.

42. Production of iron and steel in Canada during the calendar years 1907 and 1908.

43. Production of chromite in Canada during the calendar years 1907 and 1908.

44. Production of asbestos in Canada during the calendar years 1907 and 1908.

†45. Production of coal, coke, and peat in Canada during the calendar years 1907 and 1908.

46. Production òf natural gas and petroleum in Canada during the calendar years 1907 and 1908.

59. Chemical analyses of special economic importance made in the laboratories of the Department of Mines, 1906-7-8. Report on—by F. G. Wait, M.A., F.C.S. (With Appendix on the commercial methods and apparatus for the analysis of oil-shales—by H. A. Leverin, Ch. E.)

Schedule of charges for chemical analyses and assays.

†62. Mineral production of Canada, 1909. Preliminary report on—by John McLeish, B.A.

63. Summary report of Mines Branch, 1909.

67. Iron ore deposits of the Bristol mine, Pontiac county, Quebec. Bulletin No. 2—by Einar Lindeman, M.E., and Geo. C. Mackenzie, B.Sc.

†68. Recent advances in the construction of electric furnaces for the production of pig iron, steel, and zinc. Bulletin No. 3—by Eugene Haanel, Ph.D.

69. Chrysotile-asbestos: its occurrence, exploitation, milling, and uses. Report on—by Fritz Cirkel, M.E. (Second edition, enlarged.)

†71. Investigation of the peat bogs, and peat industry of Canada, 1909-10; to which is appended Mr. Alf. Larson's paper on Dr. M. Ekenberg's wet-carbonizing process: from Teknisk Tidskrift, No. 12, December 26, 1908—translation by Mr. A. v. Anrep, Jr.; also a translation of Lieut. Ekelund's pamphlet entitled 'A solution of the peat problem,' 1909, describing the Ekelund process for the manufacture of peat powder, by Harold A. Leverin, Ch.E. Bulletin No. 4—by A. v. Anrep. (Second edition, enlarged.)

82. Magnetic concentration experiments. Bulletin No. 5—by Geo. C. Mackenzie, B.Sc.

† Publications marked thus † are out of print.

83. An investigation of the coals of Canada with reference to their economic qualities: as conducted at McGill University under the authority of the Dominion Government. Report on—by J. B. Porter, E.M., D.Sc., R. J. Durley, Ma.E., and others.
  Vol.   I—Coal washing and cooking tests.
  Vol. II—Boiler and gas producer tests.
  Vol. III—(Out of print.)
    Appendix I
      Coal washing tests and diagrams.
  Vol. IV—
    Appendix II
      Boiler tests and diagrams.
  Vol. V—(Out of print.)
    Appendix III
      Producer tests and diagrams.
  Vol. VI—
    Appendix IV
      Coking tests.
    Appendix V
      Chemical tests.

†84. Gypsum deposits of the Maritime provinces of Canada—including the Magalen islands. Report on—by W. F. Jennison, M.E. (See No. 245.)

88. The mineral production of Canada, 1909. Annual report on—by John McLeish, B.A.

  NOTE.—*The following parts were separately printed and issued in advance of the Annual Report for 1909.*

  †79. Production of iron and steel in Canada during the calendar year 1909.
  †80. Production of coal and coke in Canada during the calendar year 1909.
  85. Production of cement, lime, clay products, stone, and other structural materials during the calendar year 1909.

89. Reprint of presidential address delivered before the American Peat Society at Ottawa, July 25, 1910. By Eugene Haanel, Ph.D.

90. Proceedings of conference on explosives.

92. Investigation of the explosives industry in the Dominion of Canada, 1910. Report on—by Capt. Arthur Desborough. (Second edition.)

93. Molybdenum ores of Canada. Report on—by Professor T. L. Walker, Ph.D.

100. The building and ornamental stones of Canada: Building and ornamental stones of Ontario. Report on—by Professor W. A. Parks, Ph.D.

102. Mineral production of Canada, 1910. Preliminary report on—by John McLeish, B.A.

† Publications marked thus † are out of print.

† Publications marked thus † are out of print.

NOTE.—*The following parts were separately printed and issued in advance of the Annual Report for 1911.*

NOTE.—*The following parts were separately printed and issued in advance of the Annual Report for 1912.*

† Publications marked thus † are out of print.

† Publications marked thus † are out of print.

The Division of Mineral Resources and Statistics has prepared the following lists of mine, smelter, and quarry operators: Metal mines and smelters, Coal mines, Stone quarry operators, Manufacturers of clay products, and Manufacturers of lime; copies of the lists may be obtained on application.

IN THE PRESS.

FRENCH TRANSLATIONS.

†4. Rapport de la Commission nommée pour étudier les divers procédés électro-thermiques pour la réduction des minerais de fer et la fabrication de l'acier employés en Europe—by Eugene Haanel, Ph.D. (French Edition), 1905.

26a. The mineral production of Canada, 1906. Annual report on—by John McLeish, B.A.

†28a. Summary report of Mines Branch, 1908.

56. Bituminous or oil-shales of New Brunswick and Nova Scotia; also on the oil-shale industry of Scotland. Report on—by R. W. Ells, LL.D.

81. Chrysotile-asbestos, its occurrence, exploitation, milling, and uses. Report on—by Fritz Cirkel, M.E.

100a. The building and ornamental stones of Canada: Building and ornamental stones of Ontario. Report on—by W. A. Parks, Ph.D.

149. Magnetic iron sands of Natashkwan, Saguenay county, Que. Report on—by Geo. C. Mackenzie, B.Sc.

155. The utilization of peat fuel for the production of power, being a record of experiments conducted at the Fuel Testing Station, Ottawa, 1910-11. Report on—by B. F. Haanel, B.Sc.

156. The tungsten ores of Canada. Report on—by T. L. Walker, Ph.D.

169. Pyrites in Canada: its occurrence, exploitation, dressing, and uses. Report on—by A. W. C. Wilson, Ph.D.

180. Investigation of the peat bogs, and peat industry of Canada, 1910-11. Bulletin No. 8—by A. v. Anrep.

195. Magnetite occurrences along the Central Ontario railway. Report on —by E. Lindeman, M.E.

196. Investigation of the peat bogs and peat industry of Canada, 1909-10; to which is appended Mr. Alf. Larson's paper on Dr. M. Ekenburg's wet-carbonizing process: from Teknisk Tidskrift, No. 12, December 26, 1908—translation by Mr. A. v. Anrep; also a translation of Lieut. Ekelund's pamphlet entitled "A solution of the peat problem," 1909, describing the Ekelund process for the manufacture of peat powder, by Harold A. Leverin, Ch.E. Bulletin No. 4—by A. v. Anrep. (Second Edition, enlarged.)

197. Molybdenum ores of Canada. Report on—by T. L. Walker, Ph.D.

198. Peat and lignite: their manufacture and uses in Europe. Report on— by Erik Nystrom, M.E., 1908.

202. Graphite: its properties, occurrences, refining, and uses. Report on— by Fritz Cirkel, M.E., 1907.

† Publications marked thus † are out of print.

MAPS.

†6. Magnetometric survey, vertical intensity: Calabogie mine, Bagot township, Renfrew county, Ontario—by E. Nystrom, 1904. Scale 60 feet to 1 inch. Summary report 1905. (See Map No. 249.)

†13. Magnetometric survey of the Belmont iron mines, Belmont township, Peterborough county, Ontario—by B. F. Haanel, 1905. Scale 60 feet to 1 inch. Summary report, 1906. (See Map No. 186.)

†14. Magnetometric survey of the Wilbur mine, Lavant township, Lanark county, Ontario—by B. F. Haanel, 1906. Scale 60 feet to 1 inch. Summary report, 1906.

†33. Magnetometric survey, vertical intensity: lot 1, concession VI, Mayo township, Hastings county, Ontario—by Howells Fréchette, 1909. Scale 60 feet to 1 inch. (See Maps Nos. 191 and 191A.)

†34. Magnetometric survey, vertical intensity: lots 2 and 3, concession VI, Mayo township, Hastings county, Ontario—by Howells Fréchette, 1909. Scale 60 feet to 1 inch. (See Maps Nos. 191 and 191A.)

†35. Magnetometric survey, vertical intensity: lots 10, 11, and 12, concession IX, and lots 11 and 12, concession VIII, Mayo township, Hastings county, Ontario—by Howells Fréchette, 1909. Scale 60 feet to 1 inch. (See Maps Nos. 191 and 191A.)

*36. Survey of Mer Bleue peat bog, Gloucester township, Carleton county, and Cumberland township, Russell county, Ontario—by Erik Nystrom, and A. v. Anrep. (Accompanying report No. 30.)

*37. Survey of Alfred peat bog, Alfred and Caledonia townships, Prescott county, Ontario—by Erik Nystrom and A. v. Anrep. (Accompanying report No. 30.)

*38. Survey of Welland peat bog, Wainfleet and Humberstone townships, Welland county, Ontario—by Erik Nystrom and A. v. Anrep. (Accompanying report No. 30.)

*39. Survey of Newington peat bog, Osnabruck, Roxborough, and Cornwall townships, Stormont county, Ontario—by Erik Nystrom and A. v. Anrep. (Accompanying report No. 30.)

*40. Survey of Perth peat bog, Drummond township, Lanark county, Ontario—by Erik Nystrom and A. v. Anrep. (Accompanying report No. 30.)

†41. Survey of Victoria Road peat bog, Bexley and Carden townships, Victoria county, Ontario—by Erik Nystrom and A. v. Anrep. (Accompanying report No. 30.)

*48. Magnetometric survey of Iron Crown claim at Nimpkish (Klaanch) river, Vancouver island, B.C.—by E. Lindeman. Scale 60 feet to 1 inch. (Accompanying report No. 47.)

Note.—1. Maps marked thus * are to be found only in reports.
2. Maps marked thus † have been printed independently of reports, hence can be procured separately by applicants.

Note.—1. Maps marked thus * are to be found only in reports.
2. Maps marked thus † have been printed independently of reports, hence can be procured separately by applicants.

Note.—1. Maps marked thus * are to be found only in reports.
2. Maps marked thus † have been printed independently of reports, hence can be procured separately by applicants.

Note.—1. Maps marked thus * are to be found only in reports.
    2. Maps marked thus † have been printed independently of reports, hence can be procured separately by applicants.

Note.—1. Maps marked thus * are to be found only in reports.
2. Maps marked thus † have been printed independently of reports, hence can be procured separately by applicants.

Note.—1. Maps marked thus * are to be found only in reports.
    2. Maps marked thus † have been printed independently of reports, hence can be procured separately by applicants.

†210. Location of copper smelters in Canada—by A. W. G. Wilson. Scale 197·3 miles to 1 inch. (Accompanying report No. 209.)

†215. Province of Alberta: showing properties from which samples of coal were taken for gas producer tests, Fuel Testing Division, Ottawa. (Accompanying Summary report, 1912.)

†220. Mining districts, Yukon. Scale 35 miles to 1 inch—by T. A. MacLean (Accompanying report No. 222.)

†221. Dawson mining district, Yukon, Scale 2 miles to 1 inch—by T. A. MacLean. (Accompanying report No. 222.)

*228. Index map of the Sydney coal fields, Cape Breton, N.S. (Accompanying report No. 227.)

†232. Mineral map of Canada. Scale 100 miles to 1 inch. (Accompanying report No. 230.)

†239. Index map of Canada showing gypsum occurrences. (Accompanying report No. 245.)

†240. Map showing Lower Carboniferous formation in which gypsum occurs in the Maritime provinces. Scale 100 miles to 1 inch. (Accompanying report No. 245.)

†241. Map showing relation of gypsum deposits in Northern Ontario to railway lines. Scale 100 miles to 1 inch. (Accompanying report No. 245.)

†242. Map, Grand River gypsum deposits, Ontario. Scale 4 miles to 1 inch. (Accompanying report No. 245.)

†243. Plan of Manitoba Gypsum Co.'s properties. (Accompanying report No. 245.)

†244. Map showing relation of gypsum deposits in British Columbia to railway lines and market. Scale 35 miles to 1 inch. (Accompanying report No. 245.)

†249. Magnetometric survey, Caldwell and Campbell mines, Calabogie district, Renfrew county, Ontario—by E. Lindeman, 1911. Scale 200 feet to 1 inch. (Accompanying report No. 254.)

†250. Magnetometric survey, Black Bay or Williams mine, Calabogie district, Renfrew county, Ontario—by E. Lindeman, 1911. Scale 200 feet to 1 inch. (Accompanying report No. 254.)

†251. Magnetometric survey, Bluff Point iron mine, Calabogie district, Renfrew county, Ontario—by E. Lindeman, 1911. Scale 200 feet to 1 inch. (Accompanying report No. 254.)

†252. Magnetometric survey, Culhane mine, Calabogie district, Renfrew county, Ontario—by E. Lindeman, 1911. Scale 200 feet to 1 inch. (Accompanying report No. 254.)

Note.—1. Maps marked thus * are to be found only in reports.
2. Maps marked thus † have been printed independently of reports, hence can be procured separately by applicants.

Note.—1. Maps marked thus * are to be found only in reports.
      2. Maps marked thus † have been printed independently of reports, hence can be procured separately by applicants.

†311. Magnetometric map, McPherson mine, Barachois, Cape Breton county, Nova Scotia—by A. H. A. Robinson, 1913. Scale 200 feet to 1 inch.

†312. Magnetometric map, iron ore deposits at Upper Glencoe, Inverness county, Nova Scotia—by E. Lindeman, 1913. Scale 200 feet to 1 inch.

†313. Magnetometric map, iron ore deposits at Grand Mira, Cape Breton county, Nova Scotia—by A. H. A. Robinson, 1913. Scale 200 feet to 1 inch.

Address all communications to—

DIRECTOR MINES BRANCH,
DEPARTMENT OF MINES,
SUSSEX STREET, OTTAWA

Note.—1. Maps marked thus * are to be found only in reports.
2. Maps marked thus † have been printed independently of reports, hence can be procured separately by applicants.

www.ingramcontent.com/pod-product-compliance
Lightning Source LLC
Chambersburg PA
CBHW082106210326
41599CB00033B/6610